Integration of Heterogeneous Manufacturing Machinery in Cells and Systems

With the advent of the 4th Industrial Revolution, the implementation of the nine pillars of technology has taken a firm root, especially after the post-COVID pandemic era. The integration of cyber-physical systems is one of the most important pillars that has led to the maximization of productivity, which also leads to the maximization of profits from a manufacturing system. This book discusses manufacturing enterprises, then looks at the theoretical and practical aspects of integrating these manufacturing systems using legacy and modern communication methodologies and relates them to the current level of technology readiness.

Integration of Heterogenous Manufacturing Machinery in Cells and Systems: Policies and Practices focuses on the methods covering the use of Artificial Intelligence, Augmented Reality, the Internet of Things, and cellular and physical Industrial communication. It describes the nine pillars of technology which include the Internet of Things, Cloud Computing, Autonomous, and Robotics Systems, Big Data Analytics, Augmented Reality, Cyber Security, Simulation, System integration, and Additive Manufacturing. The book highlights the methods used that cover mechanical, electrical, electronics, and computer software aspects of developing manufacturing machinery and discusses Computer-Aided Design (CAD), production planning, and manufacturing, as well as production databases with basics and semantics.

This book is an ideal reference for undergraduate, graduate, and postgraduate students of industrial, manufacturing, mechanical, and mechatronics engineering, along with professionals and general readers.

Computers in Engineering Design and Manufacturing

Series Editor: Wasim Ahmed Khan, Professor, Faculty of Mechanical Engineering, GIK (Ghulam Ishaq Khan), Institute of Engineering Sciences & Technology, Pakistan

This new series aims to acquaint readers with methods of converting legacy software and hardware used in Computer-Aided Design (CAD), Computer-Aided Manufacturing (CAM), and Computer-Aided Engineering (CAE), with or without Virtual Reality (VR) as well as current and future software modeling technology. CAD/CAM/CAE and VR technology covers Design and Manufacturing, Thermo Fluid, System Dynamics, and Control Domains will be covered in the series. The series will also address monitoring of Product Life Cycle concepts applicable to discrete products and its integration in above mentioned software packages on the World Wide Web. The concepts related to serial and parallel computing for the operation of CAD/CAM/CAE with or without VR technology at current atomic level Central Processing Unit (CPU) technology and future hardware technology based on graphene and quantum computing shall also be in the scope of the series. Current research in communication methods, both standard and proprietary at any given stage, shall also become the part of the book series at the appropriate level. The series will include books with an extensive background in mathematics, high level software languages, software modeling techniques, computer hardware technology, communication methods, and include mechanical engineering and industrial engineering domains.

Functional Reverse Engineering of Machine Tools
Edited by Wasim Ahmed Khan, Ghulam Abbas, Khalid Rahman, Ghulam Hussain, and Cedric Aimal Edwin

Integration of Heterogeneous Manufacturing Machinery in Cells and Systems: Policies and Practices
Edited by Wasim Ahmed Khan, Khalid Rehman, Ghulam Hussain, Ghulam Abbas, and Xiaoping Wang

For more information about this series, please visit: https://www.crcpress.com/Computers-in-Engineering-Design-and-Manufacturing/book-series/CRCCOMENGDES

Integration of Heterogeneous Manufacturing Machinery in Cells and Systems
Policies and Practices

Edited by
Wasim Ahmed Khan
Khalid Rehman
Ghulam Hussain
Ghulam Abbas
Xiaoping Wang

CRC Press
Taylor & Francis Group
Boca Raton London New York

CRC Press is an imprint of the
Taylor & Francis Group, an **informa** business

Designed cover image: www.shutterstock.com

MATLAB® is a trademark of The MathWorks, Inc. and is used with permission. The MathWorks does not warrant the accuracy of the text or exercises in this book. This book's use or discussion of MATLAB® software or related products does not constitute endorsement or sponsorship by The MathWorks of a particular pedagogical approach or particular use of the MATLAB® software.

First edition published [2024]
by CRC Press
2385 NW Executive Center Drive, Suite 320, Boca Raton FL 33431

and by CRC Press
4 Park Square, Milton Park, Abingdon, Oxon, OX14 4RN

CRC Press is an imprint of Taylor & Francis Group, LLC

ISBN: 978-1-032-44465-9 (hbk)
ISBN: 978-1-032-45363-7 (pbk)
ISBN: 978-1-003-37662-0 (ebk)

DOI: 10.1201/9781003376620

Typeset in Times
by MPS Limited, Dehradun

Dedication

---

*Dedicated to the Directorate General of Science
and Technology (DoST)
Department of Science and Technology & IT
Government of Khyber Pakhtunkhwa
Pakistan*

Contents

PART 1   Policies and Practices

Chapter 1   An Overview of National SME Policy – 2021 3

Muhammad Javed Afzal

PART 2 *Modern Control Theory*

PART 3   *Functional Reverse Engineering of Controller*

PART 4   4th Industrial Revolution

PART 5 *Production Philosophies and 4th Industrial Revolution*

Chapter 15 Role of Quantum Security in the Future of Smart

Danyal Maheshwari, Popenţiu-Vlădicescu Florin,
Lubna Luxmi Dhirani, Abi Waqas,
Bhawani Shankar Chowdhry, M. Mahmood Ali,
and Grigore Albeanu

Foreword

The manufacturing sector is at the forefront of technological developments and fundamental changes in the modern world today. It sets the stage for innovation and development where they are put together to shape the ways of production for our industries in the future. The book that you hold in your hands serves as a guiding light in the ever-evolving world of manufacturing, providing profound insights and thorough knowledge about the numerous components of contemporary manufacturing and the direction in which it is headed.

It provides you with an inclusive look into the essential elements of growth and development of small and medium-sized enterprises (SMEs). It helps you uncover the challenges and opportunities that lie within the realm of SMEs and their role in shaping economies worldwide. Chapter 1 of the book serves as an excellent introduction, and it prepares the readers to delve deeper into the intricacies of factors that contribute to the growth of SMEs including policy frameworks, regulatory environments, and institutional mechanisms that impact SMEs. In this chapter, the authors highlight the significance of an enabling ecosystem for the growth of SMEs and make useful recommendations for devising an effective SME policy.

The subsequent chapters enable the readers to traverse through a rich tapestry of themes that cover important aspects of modern manufacturing. From access to finance, which is extremely vital to new startups and budding SMEs, to skills development, infrastructure, and entrepreneurship – the authors provide a comprehensive understanding of the multifaceted landscape that the SMEs are compelled to navigate daily. The authors emphasize the vital roles that business development services, women entrepreneurship, and market access play in propelling SMEs towards success.

The exploration further expands to touch upon critical elements such as public procurement, institutional frameworks, and the role of organizations like SMEDA in strengthening the SME landscape. The authors meticulously examine the need for a comprehensive census of economic establishments and the importance of advocacy in creating an environment conducive to SME growth. Furthermore, they shed light on the presence of SMEDA in key regulatory arenas, demonstrating their dedication to championing the cause of SMEs.

As we progress through the chapters, the authors seamlessly transition to diverse subjects that are equally significant in today's manufacturing landscape. Chapter 3 introduces us to the realm of fault-tolerant control systems, shedding light on their architecture, fault detection and diagnosis mechanisms, and integration with feedback control. The authors delve into the complexities of dealing with unknown faults, ultimately presenting comprehensive conclusions and summaries.

Chapter 4 takes us into the world of 3-Axis WAAM Machines, exploring their design and analysis methodologies, mechanical and electrical systems, and the role of software in their functioning. The authors provide a detailed account of the development process, highlighting the integration and instrumentation aspects.

They present intriguing results and discussions that encompass factors of safety, displacement analysis, and the impact and economic analysis of these machines.

In a similar fashion, the subsequent chapters continue to unravel the mysteries and advancements within the manufacturing industry. Chapter 5 offers insights into design and analysis processes, manufacturing techniques, and concluding thoughts. Chapter 6 delves into the synthesis of machining bed kinematics, system dynamics, and controller design, with conclusive results and references.

The exploration carries on through chapters that touch upon contact-based surface estimation, digital twins, human system overview, cyber security challenges, metal forming, 3D morphology reconstruction, cyber threats, industry 4.0 advancements, and the advent of industry 5.0. Each chapter brings its unique perspective to the forefront, enriching the reader's understanding of the multifaceted nature of modern manufacturing.

As you embark on this captivating journey through the pages of this book, be prepared to witness the convergence of theoretical frameworks, empirical studies, and practical insights that will broaden the horizons of your knowledge. The knowledge within these pages is not only relevant to scholars and researchers but also to industry professionals and policymakers seeking to navigate the ever-changing landscape of manufacturing.

This book serves as a testament to the dedication, expertise, and unwavering spirit of the authors who have tirelessly curated this wealth of knowledge. Their commitment to shedding light on the complexities and opportunities within the manufacturing industry is commendable.

May this book inspire you, provoke your curiosity, and ignite your passion for the future of manufacturing. Let the words on these pages transport you to a world where innovation and progress intertwine, and where the potential for transformative change is limitless.

Prof. Dr. Iftikhar Hussain
Vice-Chancellor
University of Engineering and
Technology, Peshawar

Preface

With the advent of the 4th Industrial Revolution, the use of pillars of the industry 4.0 and Industry 5.0 has taken firm root, especially after the post covid pandemic era. "Integration" of cyber physical space is one of the most important pillars that leads to maximization of productivity, and that also leads to maximization of profits from a manufacturing system in the modern world.

Based on published work in this book series and continuing practical assignments at the Advance Manufacturing Processes Laboratory, Faculty of Mechanical Engineering, GIK Institute, the integration of manufacturing system is physically applied. The knowledge so acquired is shared with undergraduate students, graduate students, and professionals.

The book starts with the definition of small, medium, and large-scale manufacturing enterprises and then looks at the theoretical and practical aspects of integration in these manufacturing systems using legacy and modern communication and other methodologies and extends it to current technology readiness levels through a highly advanced Intelligent Manufacturing System (IMS) framework used in modern countries.

Introduction of computer numerical control (CNC) to subtractive and additive manufacturing processes is focused on, as this technology is the basis for any future integration at manufacturing cell and manufacturing systems. For legacy machinery having conventional control or numerical control (NC), the manufacturing process is included in the cyber physical space by addition of Manual Data Input Terminal (MDIT). Such hybrid manufacturing systems require communication using both legacy and modern methods. Autonomy of manufacturing process, cell, or system is maintained, with each one of these having internal and external communication controlled through handshaking and suitable authentication.

Theoretical integration of heterogenous manufacturing devices in cells and systems refers to the prevailing policies and standards related to the establishment and operation of manufacturing enterprises based on Industry 4.0 and Industry 5.0 manufacturing frameworks. The countries at advanced levels of manufacturing technologies have their own methods for this purpose as per their economic situation, avoiding extremes where the country must rely on exporting the brain and muscle power as one of the methods of earning foreign exchange instead of practicing policies that give masses food, apparel, shelter, and methods of experiencing leisure with their families together. This can be achieved through capacity building in smart manufacturing at the N11 and D8 countries.

The part on policies and practices: Theoretical and physical integration summarizes Pakistan's most recent SME Policy 2021 and its repercussions. This policy is aimed at the development and promotion of small and medium enterprises in the country. The policy is structured through a framework comprised of the key thematic areas concerning the SME stakeholders, and it also presents the mechanism of implementation of the recommendations given in the policy. This policy is typical of those underdeveloped countries structurally leaning toward that of advanced nations, and its effective implementation is normal to that of socio-political situations in these countries at the time of drafting, approving, and presenting it to industrialists for acceptance and implementation. It has bias for wealthy and upper-middle-class culture and has many tiers.

That ideal one window operation for potential SME owners remains a challenge. It ignores, to a certain extent, the understanding level of the masses for whom this policy is directed to. Furthermore, the necessary organs required to implement the policy are either too bureaucratic or are based on assumptions that lack intermediate layers of facilitation centers for developing an SME in terms of technical viability, feasibility of the proposal, finding source of finance, guidance on procurement of machinery, structure for marketing the product, and the after sales service of the product, understanding specifically the benefits of automation in terms of Industry 4.0 and Industry 5.0 pillars as well as Quantum 1.0. All these can be done through facilitation centers handling multiple SMEs at the same time. Besides that, the implementation of the SME Policy like several other national-level policies remains a challenge for bringing the real-time change on the ground that is aimed at devising the policies. Moreover, the prevailing corruption makes everything more difficult.

Chapter 2, Part 1, laments upon the physical integration in proposed framework presented in the policy response in Chapter 1. This includes technical legacy and modern methods (including Industry 4.0 and Industry 5.0 Standards) of integrating heterogenous devices to increase the productivity for much-needed products which are the necessity of Next Eleven and D8 countries and had been traditionally imported from overseas for multiple reasons including dependency of the nation on big conglomerate, multinational companies, industrialists, and landlords still working under the laws of the British Colonial System. This chapter compares the statistics available from the Pakistan Bureau of Statistics and other sources and questions why these methodologies available from the beginning of this century were not implanted in the education system to avoid the situation that this country experiences in 2023. The story seems to repeat itself from the time of freedom from the British Raj.

The privileged educated class relies more on impact factor research papers than bringing an impact. The industry academia linkages are very weak, and both sides blame each other for the problem.

A section in Chapter 2, Part 1, compares the situation in third world countries that are embroiled in political instability and extends it to Next Eleven and D8 countries who are struggling to break the shackles and have a dream of having a life as per UNDP Sustainable Development Goals by 2030, despite all the odds. It also addresses the intangible factors associated with the population of these countries that is young but is untrained and possessed by an older generation that has inertia most of the time based on nostalgia and fear that modern technology and the knowledge economy will take their children away, as with these tools children will become more independent while forgetting that cultural and religious influence binds them together.

Part 2 has the basic theme of modern control, as it is the basis of all the automation that is the hallmark of Industry 4.0 and its human factor part

Industry 5.0. More research in this area has been published in previous volumes of this book series.

Part 3 of the book is based on Functional Reverse Engineering of controller deriving theoretical and practical background from the knowledge shared in Part 2. The application of having the proposition for practical development of controllers for different types of strategic and non-strategic machines is presented in this part.

Part 4 of the book laments about the 4th Industrial Revolution and its early implementation in D8 and Next Eleven countries to catch up with the world as it happens.

Part 5 relates production philosophies with the 4th Industrial Revolution that practically tells the leader how to subside the irregularities between the centuries old philosophies and the modern Cyber Physical System (CPS) that is a hallmark of the 4th Industrial Revolution.

This book is an attempt to provide methodologies to upgrade the industry in D8 and Next Eleven countries to adopt automation, train/upskill younger generations, and bring in leisure in everybody's life through adopting the framework presented in Chapter 2, and the technology description provided in different volumes of this series.

Editors
November 2023

MATLAB® is a registered trademark of The Math Works, Inc. For product information, please contact: The Math Works, Inc.

3 Apple Hill Drive
Natick, MA 01760-2098
Tel: 508-647-7000
Fax: 508-647-7001
E-mail: info@mathworks.com
Web: http://www.mathworks.com

Acknowledgments

The authors are deeply indebted to Dr. Khalid Rahman for his help in the final review of his edited volume.

The authors would also like to acknowledge the support of Mr. Faisal Iqbal in developing the final layout of the book and the extraordinary effort in meeting the demands of the publisher to bring the book to its current state of quality and standard.

The Editors are obliged to Directorate General of Science and Technology, Department of Science and Technology and IT, Government of Khyber Pakhtunkhwa, Pakistan for their continuing efforts in supporting this endeavor.

Continuous support of GIK Institute of Engineering Sciences and Technology all the way through is deeply acknowledged.

About the Editors

Dr. Wasim A. Khan has researched and developed industrial scale subtractive and additive manufacturing machine tools, measurement, and testing machines as well as digital controllers for this equipment. He is instrumental in policies and framework development related to manufacturing of discrete products in small, medium, and large-scale manufacturing enterprises at national level. Dr. Wasim is a life member of Pakistan Engineering Council, a chartered engineer of Engineering Council, UK, and a fellow of Institution of Mechanical Engineers, UK. He is also a senior member of IEEE, U.S.A. He is currently acting as professor of Operations Research at the GIK Institute of Engineering Sciences and Technology, Pakistan.

Dr. Khalid Rahman is an associate professor in the Faculty of Mechanical Engineering at Ghulam Ishaq Khan Institute of Engineering Sciences and Technology, where he has been a faculty member since 2012. He received his B.S. in Mechanical Engineering from Ghulam Ishaq Khan Institute of Engineering Sciences and Technology, M.S. and Ph.D. degrees from Jeju National University, South Korea, in 2012. He also has industrial experience of seven years in design and manufacturing. His research interests include printing technologies, and he is currently working on Direct Ink Write and Electrohydrodynamic inkjet printing for fabrication of electronics devices and sensors and applications.

Dr. Ghulam Hussain is a professor of mechanical engineering at the University of Bahrain, Bahrain. He has rich experience of both academia and industry. He is an expert of manufacturing technologies with expertise in plasticity, sustainability, and advanced processes including 3D printing, incremental forming, friction welding and hybridization. He is the author of over 100 top-notch journal articles and an editor of several books. He is recognized among the top 10 leading researchers at the international as well as national level. He has been ranked as the world's top 2% scientist by Stanford University, U.S.A. He is the winner of many research and teaching awards/prizes. He is actively involved with several reputed international universities as a foreign expert and external researcher.

Dr. Ghulam Abbas received the B.S. degree in computer science from University of Peshawar, Pakistan, in 2003, and the M.S. degree in distributed systems and the Ph.D. degree in computer networks from the University of Liverpool, UK, in 2005 and 2010, respectively. From 2006 to 2010, he was research associate with Liverpool Hope University, UK, where he was associated with the Intelligent & Distributed Systems Laboratory. Since 2011, he has been with the Faculty of Computer Sciences & Engineering, GIK Institute of Engineering Sciences and Technology, Pakistan. He is currently working as professor and director Huawei ICT Academy. Dr. Abbas is a co-founding member of the Telecommunications and Networking Research Center at GIK Institute. He is a fellow of the Institute of Science & Technology, UK, a fellow of the British Computer Society, and a senior member of the IEEE. His research interests include computer networks and wireless and mobile communications.

Dr. Wang Xiaoping is a professor at the College of Electrical and Mechanical Engineering, Nanjing University of Aeronautics and Astronautics, a senior member of Chinese Mechanical Engineering Association, a senior member of Aeronautical Society, a guest editor of international journal *IJCAT*, an editorial board member of *American Journal of Software Engineering and Applications*, and an evaluation expert of postgraduate dissertations of the Ministry of Education. His main research areas include digital design and manufacturing and intelligent manufacturing. His current research interests are reverse engineering, aerial measurement, geodetic model reconstruction (3D reconstruction); composite material fiber laying process/ robot road planning; NC machining tool rail planning/material reduction and material increase manufacturing path optimization; CAD software system integration and algorithm; mathematical modeling; and other theoretical and technical issues related to mechanical design and manufacturing. He has executed several research projects funded by national agencies. He is also winner of three National Defense Science and Technology Progress Awards, a Provincial Award, and a Municipal Award. He has published more than 80 research articles in renowned journals and has two patents to his credit.

Contributors

Ghulam Abbas
GIK Institute of Engineering Sciences
and Technology
Topi, Pakistan

Muhammad Javed Afzal
Small and Medium Enterprises
Development Authority (SMEDA)
Ministry of Industries & Production
Pakistan

Anton. R. Ahmad
Department of Mechanical Engineering
National Taiwan University of Science
Taipei, Taiwan

Department of Mechanical Engineering
Universitas Ibn Khaldun Bogor
Bogor, Indonesia

Iftikhar Ahmad
Department of Mechanical Engineering
School of Engineering, Bahrain
Polytechnic
Isa Town, Bahrain

Grigore Albeanu
Faculty of Engineering & Informatics
Spiru Haret University
Bucharest, Romania

M. Mahmood Ali
Center for Mathematical Modeling and
Intelligent Systems for Health and
Environment (MISHE)
Atlantic Technological University
Ash Lane, Sligo, Ireland

Qasim Arain
Department of Software Engineering
MUET
Jamshoro, Sindh, Pakistan

Bhawani Shankar Chowdhry
Faculty of Computer, Electrical and
Electronics Engineering
MUET
Jamshoro, Sindh, Pakistan

Lubna Luxmi Dhirani
Department of ECE
University of Limerick
Limerick, Ireland

SFI Confirm Research Center for
Smart Manufacturing
University of Limerick
Limerick, Ireland

C. Edmondson
IMR Company Ltd
Aerodrome Business Park Rathcoole, Co.
Dublin, Ireland

Volkan Esat
Mechanical Engineering Program
Middle East Technical University –
Northern Cyprus Campus Kalkanli
Guzelyurt, Türkiye

F. O'Farell
IMR Company Ltd
Aerodrome Business Park Rathcoole, Co.
Dublin, Ireland

Abrar Farooq
Faculty of Mechanical Engineering
GIK Institute of Engineering Sciences
& Technology
Topi, Pakistan

Muhammad Umer Farooq
Institute of Mechanical and
   Manufacturing Engineering
Khwaja Fareed University of
   Engineering & IT
Rahim yar Khan, Pakistan

Popenţiu-Vlădicescu Florin
POLITEHNICA University
   Bucharest & Academy of Romanian
   Scientists
Bucharest, Romania

Adeel Ghafoor
Faculty of Mechanical Engineering
GIK Institute of Engineering Sciences
   & Technology
Topi, Pakistan

Muhammad Hammad
GIK Institute of Engineering Sciences
   and Technology
Topi, Pakistan

Ihtisham Ul Haq
Department of Mechatronics
   Engineering
University of Engineering &
   Technology
Peshawar, Pakistan

G. Hussain
Mechanical Engineering Department
University of Bahrain
Isa Town, Bahrain

Tanweer Hussain
Department of Mechanical Engineering
MUET
Jamshoro, Pakistan

Muhammad Iqbal
Institute of Computer and Software
   Engineering
Khwaja Fareed University of Engineering
   and Information Technology
Rahim Yar Khan, Pakistan

Osama Al-Jamal
Mechanical Engineering Department
University of Bahrain
Isa Town, Bahrain

S. Katyara
IMR Company Ltd
Aerodrome Business Park Rathcoole, Co.
Dublin, Ireland

Zareena Kausar
Department of Mechatronics
   Engineering
Air University Islamabad, Pakistan

Shahbaz Khan
Department of Mechatronics
   Engineering
University of Engineering &
   Technology
Peshawar, Pakistan

Wasim Ahmed Khan
GIK Institute of Engineering
   Sciences and Technology
Topi, Pakistan

Zubair Ahmad Khan
Department of Mechatronics
   Engineering
University of Engineering &
   Technology
Peshawar, Pakistan

Niaz Bahadur Khan
Mechanical Engineering Department
College of Engineering
University of Bahrain, Isa Town Campus
Isa Town, Bahrain

Muhammad Bilal Khan
Ghulam Ishaq Khan Institute of
   Engineering Science and Technology
Topi, Pakistan

Said G. Khan
Department of Mechanical Engineering
College of Engineering
University of Bahrain
Isa Town, Bahrain

Muhammad Tahir Khan
Department of Mechatronics
   Engineering
University of Engineering & Technology
Peshawar, Pakistan

Chyi-Yeu Lin
Department of Mechanical Engineering
National Taiwan University of Science
   and Technology
Taipei, Taiwan

Center for Cyber-Physical System
National Taiwan University of Science
   and Technology
Taipei, Taiwan

Taiwan Building Technology Center
National Taiwan University of Science
   and Technology
Taipei, Taiwan

P. Long
ATU Galway
Galway, Ireland

Danyal Maheshwari
eVida Research Group
University of Deusto
Bilbao, Spain

T. Morell
IMR Company Ltd
Irish Manufacturing Research Unit A
Aerodrome Business Park
   Rathcoole, Co.
Dublin, Ireland

Riaz Muhammad
Mechanical Engineering Department
College of Engineering
University of Bahrain, Isa Town
   Campus
Isa Town, Bahrain

Ali Nasir
King Fahd University of Petroleum
   and Minerals
Dhahran, Saudi Arabia

Dan Noje
University of Oradea
Oradea, Romania

Ali Akbar Shah
National Center of Robotics and
   Automation – Condition
   Monitoring Lab
MUET
Jamshoro, Sindh, Pakistan

Muhammad Faizan Shah
Faculty of Science and Technology
University of Canberra
Canberra, Australia

Syed H. Shah
Department of Mechanical Engineering
National Taiwan University of Science
   and Technology
Taipei, Taiwan

Muftooh Ur Rehman Siddiqi
Mechanical, Biomedical and
   Design Engineering
Aston University
Birmingham, UK

Radu Tarca
University of Oradea
Oradea, Romania

Xiaoping Wang
College of Mechanical and Electrical
  Engineering
Nanjing University of Aeronautics
  and Astronautics
Nanjing, China

Abi Waqas
Department of Telecommunication
  Engineering
MUET
Jamshoro, Sindh, Pakistan

Yuchun Xu
Mechanical, Biomedical and Design
  Engineering
Aston University
Birmingham, UK

Muhammad Qasim Zafar
State Key Laboratory of Tribology
Department of Mechanical Engineering
Tsinghua University
Beijing, China

Usman Zahid
Faculty of Mechanical Engineering
GIK Institute of Engineering Sciences
  & Technology
Topi, Pakistan

Fengjun Zhang
College of Mechanical and Electrical
  Engineering
Nanjing University of Aeronautics
  and Astronautics
Nanjing, China

Haiyan Zhao
State Key Laboratory of Tribology
Department of Mechanical Engineering
Tsinghua University
Beijing, China

Fatima Tu Zuhra
Institute of Mechatronics
University of Engineering & Technology
Peshawar, Pakistan

Part 1

Policies and Practices

THEORETICAL AND PRACTICAL INTEGRATION

Theoretical integration of heterogenous manufacturing devices in cells and systems refers to the prevailing policies related to the establishment and operation of manufacturing enterprises based on Industry 4.0 manufacturing frameworks. Countries have their own methods for this purpose as per their economic situation avoiding extremes where the country must rely on exporting the brain and muscle power as one of the methods of earning foreign exchange instead of practicing policies that give masses food, apparel, shelter, and methods of experiencing leisure with their families together.

The part on policies and practices: Theoretical and physical integration summarizes Pakistan's most recent Small and Medium Enterprises policy and its repercussion. This policy is typical of those of underdeveloped countries structurally leaning towards that of advanced nations and its effective implementation is normal to that of socio-political situations in these countries at the time of drafting, approving, and presenting it to the industrialists for acceptance and implementation. It has very high bias of upper-middle-class culture and altogether ignores the understanding level of the masses for whom this policy is directed to. Furthermore, the necessary organs required to implement the policy are either missing or are based on naivete assumptions.

A brief policy response is provided by the editors towards the end of Chapter 1, Part 1.

Chapter 2, Part 1, laments on the physical integration in proposed frameworks presented in the policy response in Chapter 1. This includes technical legacy and modern methods (including Industry 4.0 Standards) of integrating heterogenous

DOI: 10.1201/9781003376620-1

devices to increase the productivity for much-needed products which are the necessity of this country and had been traditionally imported from overseas for multiple reasons including dependency of the nation on big conglomerate, multinational companies and landlords still working under the laws of British Colonial System. This chapter compares the statistics available from the Pakistan Bureau of Statistics and questions why these methodologies available from the beginning of this century were not implanted in the education system to avoid the situation that this country experiences in 2023. The story seems to repeat itself from the time of freedom of the British Raj.

A section in Chapter 2, Part 1, compares the situation in third-world countries that are embroiled in political instability and extends it to Next Eleven Countries that are struggling to break the shackle and have a dream of having a life as per UNDP Sustainable Development Goals by 2030 despite all the odds. It also addresses the intangible factors associated with the population of these countries that is young but is untrained and possessed by an older generation that has inertia most of the time based on nostalgia and fear that the modern technology and knowledge economy will take their children away as with these tools children will become more independent while forgetting that cultural and religious influence binds them together.

1 An Overview of National SME Policy – 2021

Muhammad Javed Afzal

1.1 CONSTRAINTS TO SME GROWTH

In Pakistan, the SME sector comprises a variety of business sectors such as manufacturing, services, and trading enterprises. The SME policy highlights several issues that hinder their growth, including lack of access to formal finance and resources, shortage of skilled labor and technology, and insufficient capacity for R&D. As a result, most SMEs remain stuck in the lower end of the value chains and unable to scale up and diversify. Furthermore, small businesses are burdened with high costs of regulatory and tax compliance, which often leads them to stay informal or unregistered. Pakistan's export share is mostly comprised of low-value-added products, and the policy and business environment have not been conducive to SME growth. Additionally, the infrastructure for SMEs, particularly in the important clusters of the manufacturing sector, is inadequate and has disparities between provinces.

1.2 POLICY FRAMEWORK

The framework of National SME Policy 2021 comprises four key domains including the following.

1.2.1 MACRO POLICY AND REGULATORY ENVIRONMENT

The Government's National SME Policy 2021 aims to implement measures that support the growth of SMEs by creating a stable macro environment. The policy proposes a reform strategy to reduce the regulatory burden of SMEs, which affects the cost of doing business. The reforms will address the issues in respective domains of policy, regulatory, and compliance.

1.2.2 SUPPLY-SIDE CHALLENGES

According to the Policy, SMEs still face challenges in accessing essential resources that are crucial for their growth and competitiveness. These resources include affordable and high-quality business development services, improved

DOI: 10.1201/9781003376620-2

access to credit, better human resources, innovation, and opportunities for entrepreneurship, as well as sustainable infrastructure. To address the supply-side challenges, the Policy recommends sector-specific interventions in high-growth economic sectors.

1.2.3   DEMAND-SIDE CHALLENGES

The Policy recommends support for SMEs to improve access to both local and export market and support for increasing the opportunities for demand of the produce and services of SMEs.

1.2.4   INSTITUTIONAL MECHANISM

National SME Policy emphasizes on reinforcing the voice of SMEs and also gives an institutional mechanism for the application of policy recommendations.

1.2.5   KEY RECOMMENDATIONS IN SME POLICY 2021

The following are the key areas of the recommendations in National SME Policy 2021.

1.3   SME DEFINITION

Defining SME categories is essential to identify enterprises that require targeted support interventions. These interventions can include regulatory exemptions, preferential tax treatment, subsidized BDS, and access to public procurement and credit. Globally, the common parameters used for defining SMEs are the number of employees, the value of assets, and the turnover. In Pakistan, the SME Policy 2021 outlines the following definition of SMEs to be used across the country (Table 1.1).

TABLE 1.1
Criteria for Definition of SME

Category of SME	Criteria for Definition of SME (Annual Sales Turnover)
Small Enterprise (SE)	Up to PKR 150 Million
Medium Enterprise (ME)	Above PKR 150 Million to PKR 800 Million
Start-up	A small enterprise or medium enterprise upto 5 years old

1.4   REGULATORY AND TAX ENVIRONMENT

The regulations related to SMEs in Pakistan cover various aspects such as licensing, labor laws, market entry and exit procedures, insolvency, contract enforcement, and tax rates. However, excessive regulations increase the cost of doing business and discourage full disclosure, which hinders the growth of SMEs. The SME Policy 2021 proposes a set of the following measures to improve the business environment in Pakistan.

- The government will establish a registry of rules and regulations to improve the regulatory system. Rules and regulations not part of this registry will not be imposed on enterprises, and regulatory impact assessment will be part of the approval mechanism for new rules and regulations.
- Inspections into business premises will be reduced and a smaller sub-sample of businesses will be inspected. Inspectors of the regulatory bodies will no longer have discretionary powers to seal the premises or impose immediate fines, and all fees and taxes to be paid through e-challans.
- E-Inspection Portal will be established for authorizing, scheduling, and validation, accessible to line departments and SMEs, both at federal and provincial levels.
- A "No NOC/Self Declaration and Time-Bound Approval" regime will be introduced for new SMEs, Expansion of business, and Balancing, Modernization, and Replacement (BMR). Time-bound approval system to be introduced.
- The Federal Board of Revenue (FBR) will launch a facilitative SME tax regime to benefit selected service sectors such as IT and ITES.
- FBR, Federal, and Provincial Governments will consider consolidation, harmonization, and amalgamation of taxes and levies rates and collection systems at provincial and federal levels.
- FBR will consider gradually reducing Withholding Tax with a corresponding increase in formalization and Sales/Income Tax receipts.
- Import tariffs on raw materials and machinery will be rationalized, with a special focus on the SMEs.
- SMEs may be prioritized for sales tax refund and duty drawback refund.
- The government will establish bonded warehouses for import of important inputs and strict penalties on cartelization of commercial importers of raw materials will be enforced by the Competition Commission of Pakistan.
- The government will introduce a presumptive tax option for a category of small and medium enterprises.
- Small companies will pay 20% corporate tax, as compared to 29% for large companies.
- Incentive schemes will be designed to channelize savings into equity finance.

- The government will reduce the frequency of tax payments from 47 to 34, and the process will be refined through a system of e-challans and e-payments.
- The Tax Ombudsman will be strengthened, and a separate window for SMEs will be established, with a time limit set for disposal of all complaints.
- SECP and IPO Pakistan will facilitate registration of businesses by creating an online database of names, logos, and patents, respectively, that can be checked in real-time by firms when selecting a name.
- The Provincial departments of industries and SECP will develop a single portal for online registration that automatically registers enterprises with selected organizations such as EOBI, provincial social security institutes, and department of labor.
- A single unique identifier will be issued to all registered firms that can be used with all government departments to pull up relevant data electronically.
- IT may be used to the extent possible for simplifying and expediting interface with the government, including invoice systems with tax record-keeping.

1.5   ACCESS TO FINANCE

Access to credit is a significant challenge faced by SMEs in Pakistan. A majority of SMEs rely on self-financing, informal financing, or personal savings. This limited financial capacity hinders their business operations and limits their ability to expand. Even though the regulatory framework for SMEs to raise funds from the capital market exists, the uptake by SMEs is minimal. To improve SMEs' access to finance, the SME Policy suggests the recommendations, including:

- Encouraging banks to lend to SMEs equitably under the SAAF scheme through SBP.
- Considering relaxing the age limit to 60 years for SMEs to become eligible for the Kamyab Jawan financing scheme.
- Enhancing credit limits for micro and small enterprises.
- Reforming the VC regime to facilitate growth of SMEs in the IT and ITES sector.
- Facilitating SMEs to avail lower interest rate schemes announced by SBP.
- Ensuring implementation of the National Financial Inclusion Strategy (NFIS) 2023.
- Providing quarterly progress reports to SMEDA on financing to SMEs and publishing them in the State Bank's Quarterly SME Finance Review.
- Reviewing key regulations by SBP, relating to SME financing and reducing the cost of accessing and providing capital.
- Disseminating information about the movable asset registry to ensure its uptake by commercial lending institutions.

- SBP to streamlining the process of account opening and documentation required by commercial banks.
- Facilitation by the commercial banks to open back-to-back Letter of Credit (LC) for exporting firms to use their receivables as collateral to secure short-term financing.
- Operationalizing the Pakistan Credit Guarantee Company and incentivizing banks by offering different risk coverage by SBP.
- Strengthening SME data collection and initiating the process of gathering industry-level and firm-level data required for credit scoring and risk assessment.
- Ascertaining the viability of establishing a credit risk rating and assessment agency that can provide consulting services to banks and lending institutions, as well as SMEs.
- Conducting value chain studies on key clusters to identify gaps and opportunities.
- Conducting a study and viability analysis of using innovative blended finance, such as Islamic products for enhancing the flow of resources to SMEs.
- Ensuring better utilization of SBP's refinance and credit guarantee scheme for women entrepreneurs.
- Carrying out awareness campaigns for SMEs, by Pakistan Stock Exchange (PSX), regarding raising finance through the capital market and Facilitating and incentivizing SMEs to enlist on the GEM Board of the PSX.
- Undertaking capacity-building initiatives for SMEs on Accounting and Bookkeeping in consultation with ICAP, ICMAP, SBP, NIBAF, and other institutions.
- Reviewing options (privatization and/or restructuring) for the SME Bank by the Government of Pakistan, given that the Bank remains the only FI focused exclusively on SMEs.
- Establishing a committee to devise a strategy for expanding SME financing in the country.

1.6   SKILLS, HUMAN RESOURCES, AND TECHNOLOGY

The 2021 SME Policy recognizes the lack of skills and training as a major obstacle to enterprise growth. To address this issue and enhance the capacity of HR, skills, and technology, the National SME Policy has recommendations as follows:

- Ensure implementation of the National TVET Policy.
- Skills mapping for critical SME sectors and develop future programs of skill training accordingly.
- Include SMEDA representation on the boards of NAVTTC and provincial TEVTAs to voice SME concerns.
- Collaborate with provincial TEVTAs to streamline overlaps and regulatory burdens for private-sector investments in skill development.

- Evaluate the viability of replicating the Punjab Skills Development Fund (PSDF) model in coordination with provinces.
- Identify SME sector-specific training requirements by SMEDA and share them with provincial TEVTAs.
- Develop a "Soft Skills" training module to be integrated into NAVTTC and provincial TEVTAs courses.
- Establish a partnership framework between provincial TEVTAs and key SME sector associations to develop industry-relevant courses.
- SMEDA to identify the necessary skills for updated technology used, in collaboration with TUSDEC, and communicate them to NAVTTC and provincial TEVTAs to create new programs.
- Assist SMEs in acquiring appropriate technologies to boost productivity, quality, and competitiveness.
- Skilling Pakistan and provincial TEVTAs to develop a sustainable labor market data management model.
- SMEDA to advocate for increasing women-centric training programs.

1.7 INFRASTRUCTURE

Various studies reveal that a lack of infrastructure facilities affects the productivity and competitiveness of SMEs. SME Policy 2021 has the following key recommendations to improve the infrastructure:

- SMEs will have access to plug-and-play infrastructure on a lease basis.
- Existing industrial estates, EPZs, and SEZs will allocate space for SMEs.
- Government-owned land will be used to establish new SME industrial estates.
- SMEs will be allotted adequate space in SEZs under CPEC and other industrial estates.
- The Power Division will consider shifting manufacturing SMEs from commercial electricity connections to Industrial electricity connections (Bl, B2 only) with access to lower tariffs.
- Support for off-grid clean energy solutions will be provided to enable SMEs to access affordable energy.
- The Federal Government will assist provincial Governments in developing spatial mapping, similar to the Punjab province, as a central planning tool for all infrastructure development decisions that take into account the connectivity needs of SME clusters.
- Farm-to-market roads will be constructed to connect small and rural enterprises to major roads.
- Commercial electricity consumers will receive a seasonal discount on incremental consumption as part of the PM's Industrial Support Package for power.

1.8   ENTREPRENEURSHIP, INNOVATION, AND INCUBATION

A conducive entrepreneurship ecosystem encourages youth to become entrepreneurs and can support start-ups to scale up into larger businesses. The National SME Policy 2021 includes the following recommendations to strengthen a culture of entrepreneurship in the country:

- Encourage provinces to adopt the more flexible Limited Liability Partnership Act 2017 as a partnership registration model.
- Establish legal frameworks for crowdfunding, venture capital funds, private equity funds, etc., and develop suitable rules for their registration with the SECP.
- Implement legal reforms to simplify business closure, including restructuring of debt options.
- Implementation of the Corporate Rehabilitation Act, 2018.
- Expand the regulatory sandbox approach to monitor and regulate evolving and developing new businesses such as fintech.
- Evaluate and scale up Business Incubators with reserved seats for women.
- Initiate programs in schools to teach entrepreneurship and develop such skills among students.
- Provide training to entrepreneurship boot camps for youth to help them access the Government's youth entrepreneurship schemes/programs.

1.9   BUSINESS DEVELOPMENT SERVICES

Non-Financial Advisory Services (NFAS), also known as Business Development Services (BDS), are services that provide business advice and skills transfer. They include consultancy and advisory services, training, marketing assistance, technology development and transfer, and promotion of business linkages. To support the growth of private-sector businesses, the National SME Policy recommends the following:

- Conducting a needs assessment survey for BDS by SMEDA.
- Linking SMEs with Business Development Service Providers (BDSPs) and sharing the cost of BDS services through funding from SMEDA.
- Creating an online ranking system for SMEs to share feedback on BDSPs.
- In the long run, registering BDSPs by SMEDA and providing them access to the database of registered SMEs, negotiating the cost of services, and overseeing the quality and efficiency of the BDS provided.
- Identifying high-potential sectors and taking initiatives for growth (such as garments, light engineering, leather, agro-food, IT and ITES, auto, tourism, pottery, surgical, furniture, pharmaceuticals, electronics, etc.). The following SME categories will be given priority:
  - Vendors of large-size businesses.
  - SME exporters.
  - SMEs in peripheral regions.

1.10 WOMEN ENTREPRENEURSHIP DEVELOPMENT

Encouraging women's entrepreneurship has been shown to stimulate economic growth and create jobs. The National SME Policy 2021 recognizes this and recommends preferential treatment for women-owned businesses within the SME regulatory and business support arenas, with the following recommendations:

- Establish a Consultative Group, led by MoIP, to support Women-owned SMEs with input from public- and private-sector stakeholders.
- Implement the "Banking on Equality" policy of the State Bank of Pakistan.
- Simplify taxation procedures and reduce tax rates for women-owned SMEs.
- Create a virtual one-window facility for women entrepreneurs to access information.
- Establish women-friendly work environments, such as women business centers, display facilities, and emporiums.
- Develop specialized training programs for women in areas such as starting and managing a business, accounting and bookkeeping, fulfilling tax liabilities, marketing, and digitization.
- Build linkages with domestic and international markets.

1.11 MARKET ACCESS

National SME Policy 2021 recommends the following measures to improve market access for SMEs:

- To address the issue of costly access to major international cities and markets, TDAP will organize and facilitate small firms and women entrepreneurs in international fairs and exhibitions at subsidized rates.
- Various funding options will be explored to support SME participation in international trade fairs and exhibitions, including the funding from EDF, SME Development Fund, and development partner networks.
- To improve local market access for small businesses and cottage industries located in remote areas, to have access to regular trade fairs and exhibitions in all major cities of Pakistan.
- Permanent emporiums will be established in major cities to showcase the arts, crafts, and cuisine of all provinces and regions of the country.
- The National Product Standards will be developed, strengthened, and implemented to ensure quality products and services.
- The E-Commerce Policy of Pakistan 2019 will be implemented to incentivize the establishment of "Online Market Place/Digital Platforms" to facilitate SMEs in accessing both domestic and international markets.
- SMEs will be encouraged to participate in international e-commerce, and the State Bank will expedite the approval of applications of local investors to operate payment gateways.

1.12   PUBLIC PROCUREMENT

The National SME Policy 2021 recommends using public procurement as a tool to support market access opportunities for SMEs. Government procurement, which follows a specific set of rules and procedures, can act as a major client for SMEs and encourage formalization of informal entities. The policy proposes the following recommendations to provide SMEs with a level playing field and to compete for public contracts:

- The Federal PPRA should develop a mechanism to increase SMEs' share in public procurement.
- Public procurement regulatory authorities at federal and provincial levels should revise procurement rules to facilitate greater SME participation in public procurement.
- PPRA (Federal and Provincial) should allocate quotas for SMEs.
- SMEDA should support the establishment of procurement support units at the chambers of commerce and industry and trade associations to help cope with capacity issues of SMEs.

1.12.1   INSTITUTIONAL FRAMEWORK

Implementation of the National SME Policy 2021 involves multiple departments concerned. To ensure the implementation of the National SME Policy, there is the following mechanism.

1.13   NATIONAL COORDINATION COMMITTEE (NCC) ON SMES DEVELOPMENT

The Government constituted a National Coordination Committee on SMEs Development to lead SME development and ensure effective implementation of the National SME Policy 2021. The structure of the NCC on SMEs Development is as follows:

i. Minister for Industries and Production – Convener
ii. Secretary, Ministry of Industries and Production
iii. Secretary, Commerce Division
iv. Secretary, Finance Division
v. Secretary, Power Division
vi. Secretary, Petroleum Division
vii. Secretary, Ministry of Law and Justice
viii. Secretary, BOI
ix. Chief Secretaries of the Provincial Governments
x. Chairman, FBR
xi. Chairman, BOI
xii. Deputy Governor, SBP

xiii. Chairman, SECP
xiv. CEO, TDAP
 xv. CEO, SMEDA
xvi. One (1) Representative of SME sector from each province/key sectors
      Representatives from the SME Sector (All provinces, Gilgit Baltistan
      and Azad Jammu and Kashmir).

The NCC on SMEs Development will be assisted by the Provincial Working
Groups in the provinces. Working Groups are to support provincial implementation
and to bring up issues that SMEs face in the provinces that need to be addressed
at the federal level.

1.14   SMEDA'S INSTITUTIONAL STRENGTHENING

SMEDA is the primary SME development agency of the Federal Government,
responsible for promoting SME growth in the country. SMEDA's mandate, as
outlined in the SMEDA Ordinance 2002, enables the agency to undertake all
necessary actions to encourage and facilitate the growth and development of SMEs.
To effectively implement the National SME Policy 2021 and fulfill its mandate,
SMEDA will be strengthened through the following institutional capacity-building
initiatives and increased resource provision by the Government of Pakistan:

* Enhancing fund management capacity.
* Strengthening policy and strategy management.
* Providing business development services.
* Building entrepreneurship and skills expertise.
* Developing expertise in emerging technologies, high-growth sectors (such
  as IT and ITES), and the knowledge economy.
* Strengthening monitoring and evaluation capabilities.

1.15   SME REGISTRATION PORTAL

The limited availability of consolidated data on SMEs and their characteristics
hinders the government's ability to develop evidence-based policies and programs
and provide targeted incentives and relief where needed. To address this, SMEDA
will establish an SME Registration Portal (SMERP) that will be integrated with
other business databases and portals for verification, credit assessment, and market
linkage. SMERP will facilitate voluntary registration of sole proprietors, firms, and
companies according to SME definition to create a centralized database of SMEs in
the country. Additionally, SMEDA will create an online formal consultation forum
with SMEs on regulatory topics, which will also serve as a platform for accessing
information on business regulations, services, government incentives, identifying
potential partners for larger orders, coordinating joint investments, finding suppliers
or large firms seeking SME suppliers, and ranking and providing feedback on
business service providers. SMEs can also apply for SME Size Certificate through
the SME Registration Portal.

1.16   SME FUND

SMEDA will create an SME Fund to guarantee sufficient funding for its operations and SME growth initiatives. The following projects will be undertaken by the fund.

- SME technology upgrade.
- Development of SME credit market.
- Support for SMEs in BDS, technology and product development, certification, access to market, training, efficiency enhancement, and so on.
- Provide equity financing, startup capital, as well as other technology-based financial options.
- Assistance in creating an entrepreneurial ecosystem.
- Assisting SMEs with industry research, marketing, and exhibitions.

1.17   CENSUS OF ECONOMIC ESTABLISHMENTS/SME CENSUS

The Census of Economic Establishments will be conducted by the Pakistan Bureau of Statistics (PBS). The information gathered will be used to populate the SMEDA National Database on SMEs.

1.18   ADVOCACY

As the main organization for SME development, SMEDA will play an active part in advocacy of SME and coordination of SME-related initiatives throughout the country to foster an environment conducive to new business formation, entrepreneurship, and demand-based BDS for SMEs.

1.19   PRESENCE OF SMEDA IN KEY REGULATORY ARENAS

SMEDA will be represented in trade, fiscal, and taxes policy organizations such as the FBR, SBP, and Ministry of Commerce, as well as other agencies to assure that the SME voice is evident at pivotal moments of policy making based on the requirements of smaller businesses. SMEDA will also be represented on the boards/committees of all small industries provincial departments/entities, the boards of provincial TEVTAs and NAVTTC, and skills sector councils, the SBP Committee on financing of SME, and the programs and platforms of international development partners.

1.20   DISCUSSION AND CONCLUSION

Small- and medium-sized enterprises (SMEs) play a significant role in the economic development of a nation. Governments worldwide prioritize the development of SMEs to boost their national progress. In Pakistan, SMEs contribute to the economy through job creation, human resource development, value addition, and innovation. To fully exploit the potential of the SME sector and lead to economic growth, the Pakistani government has prioritized SME development through this National SME

Policy 2021. This policy aims to make Pakistani SMEs globally competitive and innovative, potentially creating high-value jobs. It focuses on key areas such as access to finance, regulations and tax regimes, skills, industrial infrastructure, and promoting entrepreneurship. The policy has two pillars: reforming the policy and regulatory environment and addressing SME market constraints, both on the demand and supply sides.

The policy focuses to create a business-friendly environment, simplify regulations, and institute a regime that allows easy entry and exit for business start-ups. On the supply side, the policy focuses on fiscal and monetary incentives, SME facilitation, entrepreneurship and innovation, credit and skills, and infrastructure provision. On the demand side, it focuses on market access, women entrepreneurship, and access of SMEs to public procurement. The policy recommends a unified definition of SMEs to be followed by all stakeholders and a voluntary registration process for SMEs to support the building of a National Database of SMEs. The institutional framework for implementation of the policy includes a National Coordination Committee (NCC) on SME Development, strengthening of the Small and Medium Enterprises Development Authority (SMEDA), an SME Registration Portal, a census of economic establishments, and advocacy for SMEs.

The SME Policy 2021 thus provides a framework and action plan for sustained efforts to achieve a globally competitive SME sector that can effectively contribute to reviving the economy of Pakistan. It is hoped that with proper implementation of this policy, the SMEs in Pakistan will leverage these policy interventions for increased contribution to GDP, employment, and exports over time.

REFERENCES

https://moip.gov.pk/userfiles1/file/SME%20Policy%20-%202021.pdf
Extrapolated from data of Census of Economic Establishments 1988 and 2005.
Labor Force Survey2017-18.
Pakistan Economic Survey2010-11.
Pakistan Economic Survey2009-10.
Quarterly SME Finance Review – State Bank of Pakistan June 2021.
Census of economic establishments 2005, Pakistan Bureau of Statistics.

2 IMS Framework for N11 and D8 Countries

Muhammad Hammad, Ghulam Abbas, and Wasim Ahmed Khan

2.1 INTRODUCTION

Industry 4.0 Pillars lead to cyber physical genetic integration while Industry 5.0 pillars consider intangible human factors. Industry 4.0 pillars are as follows:

1. Autonomous Robots
2. Simulations/Augmented Reality/Virtual Reality
3. Horizontal and Vertical System Integration
4. Industrial Internet of Things
5. Cloud Computing
6. Cybersecurity
7. Additive Manufacturing
8. Big Data Analytics
9. Artificial Intelligence
10. Blockchain
11. Molecular Biology/Genetics

Pillars of Industry 5.0:

1. All human factors (implemented or not implemented in the industry)

While consideration is given to as many pillars of Industry 4.0 and Industry 5.0 as possible, Horizontal and Vertical Integration in Cyber Physical Genetics space is given primary importance. Integration, both in terms of theory and physical implementation, is addressed. The SME policy of Pakistan is taken on a theoretical basis, while data communication standards are taken as the physical part of the study.

Based on Schwab's conception, it is projected that this is the best chance for N11 and D8 countries to catch up with the world in the technology arena and do not miss the boat again. Such chances come up in centuries.

Two factors are considered important here: (1) Data Communication protocols and (2) IMS and IMIS frameworks for SMEs and startups of N11 and D8 Countries.

DOI: 10.1201/9781003376620-3

These are described in the following sections.

Frameworks for Intelligent Manufacturing System and Intelligent Manufacturing Implementation System:

Intelligent manufacturing (IM) or advanced manufacturing is a comprehensive approach to shaping conventional manufacturing processes into highly advanced, efficient, flexible, and connected operations. In IM, the complete manufacturing system is interconnected in such a way to provide seamless collaboration, information sharing, and optimization from the suppliers to customers [1].

Intelligent Manufacturing System (IMS):

Product life cycle (P), Industry 4.0 (I), and Design and Manufacturing (D) are the three axes that generate the framework for Intelligent Manufacturing System (IMS), as shown in Figure 2.1. Product life cycle (P) refers to the complete journey passing through the sequence of stages from initial design to eventual reuse/recycle. Industry 4.0 (I) contains the basic pillars that cover the key technologies and concepts that help to define the fourth industrial revolution. Design and manufacturing (D) encompasses different stages of design activities and manufacturing processes involved in the creation and production of goods [2].

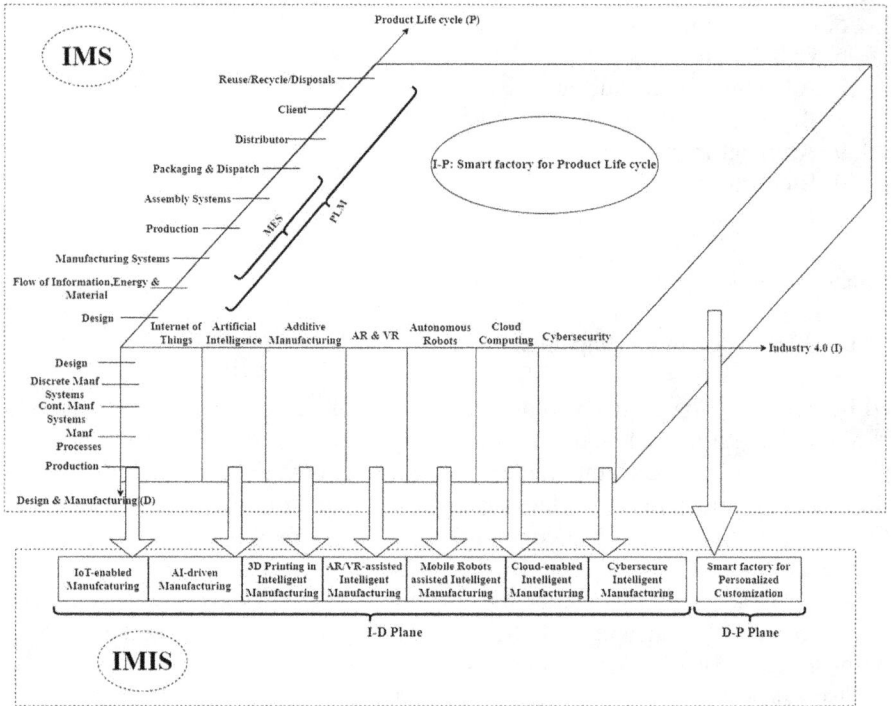

FIGURE 2.1   Framework for IMS and IMIS.

The stages of product life cycle (P) are as follows [3]:

- Design: This stage covers conceptualizing, specifications, features, functionality, and creating the product. IMS leverages technologies like computer-aided design (CAD) and simulation tools to generate the digital model and optimize the product design.
- Flow of Information, Energy, and Material: There is a concentration on efficient and seamless required information in this stage. The required data related to design specifications, material requirements, and quality control parameters are shared across different departments.
- Manufacturing Systems: In this stage, the raw material is transferred into finished products using advanced manufacturing systems and technologies including automation, robotics, and digital control systems. By doing so, productivity and quality can be enhanced while ensuring real-time monitoring and decision-making.
- Production: In this stage, the actual manufacturing of the product according to design specification takes place. IMS leverages automation, data analytics, and connectivity to identify any production issue or bottleneck to optimize the production workflows and monitor key performance indicators.
- Assembly Systems: This stage integrates different components or sub-assemblies to create the final product. IMS used robotics automation and computer vision to increase the speed, flexibility, and accuracy in the assembly process.
- Packaging and Dispatch: After the product is assembled correctly, it requires to be packed securely for the distribution and transportation stage. IMS used automation in packaging processes to lessen the waste and improve logistics.
- Distributor: This stage involves the transportation of packed products from the manufacturing department to the intended distributors or retailers. IMS facilitates data analytics and connectivity to manage inventory, demand forecasting, and supply chain optimization to ensure timely delivery and efficient distribution.
- Client: This stage involves the purchase and usage of the product by the customer. Customer satisfaction can be improved by incorporating technologies like IoT-enabled devices and customer feedback systems.
- Reuse/Recycle/Disposal: This stage involves the management of the product at its end-of-life by focusing on sustainable practices. The following processes such as disassembly for recycling and refurbishment for reuse are used for the purpose.

Manufacturing Execution System (MES): MES has a range from manufacturing systems to packaging and dispatch. Product lifecycle management (PLM): It has a range from the flow of information, energy, and material to the reuse/recycle/disposal as shown in Figure 2.1.

The basic pillars of Industry 4.0 (I) considered here are as follows [4]:

- Internet of Things (IoT): IoT is the network of interconnected physical devices embedded with sensors and actuators. IoT plays a significant role in integrating machines, devices, and systems to enable real-time data collection.
- Artificial Intelligence (AI): It involves the development of intelligent machines that can perform tasks autonomously that typically require human intelligence. AI used different technologies such as machine learning, deep learning, and predictive analysis for the required purposes.
- Additive Manufacturing: Also called 3D printing, it involves the process of producing the product by adding materials in layers. It enables the production of intricate, customized, and lightweight parts with greater design flexibilities, rapid prototyping, and customization.
- Augmented Reality (AR) and Virtual Reality (VR): It involves the creation of an entirely virtual environment by providing immersive and interactive experiences by overlaying digital information onto the physical world.
- Autonomous Robots: These are machines that can perform tasks with little or no human involvement. Various operations can be performed by autonomous robots such as material handling, inspection, assembly, and logistics.
- Cloud Computing: It involves the delivery of on-demand computing resources, such as storage, processing power, and software applications, over the internet. It promotes secure data storage, real-time data analysis, and collaboration among stakeholders.
- Cybersecurity: It helps to protect the data, systems, and networks from unauthorized access, breaches, and cyberattacks. Different robust and efficient cybersecurity measures are used to secure the system.

The design and manufacturing (D) axis consists of the following stages [5–7]:

- Design: It is the initial phase that enables conceptualizing and creates a product's specifications and features. It is a critical stage, as it lays the foundation for the manufacturing processes.
- Discrete Manufacturing Systems: It encompasses the production of distinct items that can be counted. It has the complex assembly processes to create the final product.
- Continuous Manufacturing Systems: It involves the production of goods in uninterrupted and continuous flow. The common industries using these types of systems are chemicals, food processing, and pharmaceuticals. The use of automated processes to ensure efficient and continuous manufacturing makes these systems more prominent.
- Manufacturing Processes: It contains the specific techniques and processes to convert raw materials into finished products.
- Production: It covers the overall management and execution of manufacturing processes. It encompasses the coordination of resources, labor, and equipment to ensure the planned manufacturing activities to be carried out effectively.

Intelligent Manufacturing Implementation System (IMIS):

For implementation, any two axes of the IMS combine to form the plane of IMIS. For example, the 'I' and 'P' axes merge to form the I-P plane, the 'I' and 'D' axes collectively generate the I-D plane, and similarly, the 'D' and 'P' axes combine to form the D-P plane.

I-P plane: This plane is related to smart factory for product life cycle. It is a highly advanced and digitalized manufacturing ecosystem where advanced technologies are integrated throughout the entire product life cycle. It contains the following stages [8]:

- Design: Advanced tools are used to enable digital models and simulations that ultimately help manufacturers to analyze and validate design before actual production. It minimizes errors and speeds up the design process.
- Production Planning: Different advanced analytics and machine learning algorithms are used to optimize production planning. Advanced tools enable inventory management, demand forecasting, scheduling, and resource allocation to lessen waste and ensure efficient production.
- Manufacturing Execution: This is an IoT-enabled manufacturing system where machines, sensors, and production lines are engaged in a way to provide real-time data. This data enables quality control, predictive maintenance, and real-time optimization to enhance the efficiency of production processes.
- Supply Chain Integration: It enhances collaboration with suppliers, customers, and distributors by leveraging data exchange and connectivity.
- Product Quality Assurance: Advanced quality control systems (AQCS) are used to monitor and ensure the product quality throughout the entire manufacturing process. AQCS may include machine vision, sensors, and real-time data analysis.
- Postproduction Services: Smart factory utilizes remote diagnostics, real-time monitoring, and predictive maintenance to ensure the proactive support and timely service of the products in the field. It improves the long-term customer experience and loyalty.
- Continuous Improvement: Data analytics and machine learning algorithms diagnose the areas for improvements in the product life cycle. It will lead to cost reduction, process optimization, and innovation in product design and manufacturing.

I-D plane: This plane is related to the coordination of Industry 4.0 pillars with the design and manufacturing to implement intelligent manufacturing. It contains the following fields:

1. IoT-enabled Manufacturing: It involves the integration of IoT in manufacturing industries to enable intelligent manufacturing. It covers the following domains [9,10]:
   - Devices and Sensors Integration: It contains the network of various sensors and devices within the manufacturing ecosystem. These sensors are embedded with machines, equipment, or production lines to detect

anomalies, monitor performance, and optimize operations. Sensors may have the capacity to collect data regarding temperature, humidity, pressure, vibration, and machine status.

- Data Acquisition and Analysis: The sensors used in the IoT system collect a large amount of data related to energy usage, machine performance, production metrics, environmental conditions, and many more. This data needs to be analyzed using IoT platforms to perform actionable insights, identify patterns, and make the data-driven decision to optimize the processes.
- Real-time Monitoring and Control: One of the benefits of the IoT-enabled system is to provide real-time monitoring and control capabilities. Real-time data helps manufacturers to make proactive decisions, predictive maintenance, and optimization of the system under consideration.
- Energy Efficiency and Sustainability: Smart sensors collect data related to energy consumption, enabling manufacturers to identify energy-saving opportunities to increase energy efficiency. It will ultimately reduce costs and enhance sustainability.
- Worker safety and productivity: IoT-enabled systems use wearable devices, safety equipment, and real-time monitoring systems that enable them to detect any hazardous situations, including the tracing of workers in case of alarming conditions. It also enables task automation and workflow optimization to increase worker productivity.

2. AI-driven Manufacturing: It refers to the integration of AI and machine learning algorithms to enable intelligent manufacturing with the aim to optimize and automate various aspects of manufacturing processes. It covers the following features [11]:

- Predictive Analytics: AI-driven manufacturing facilitates predictive analytics as algorithms analyze large volumes of data to forecast outcomes and identify patterns from the historical data.
- Intelligent Process Automation: AI algorithms can automate repetitive tasks to enable faster, accurate, and optimal production processes. Repetitive tasks may include data entry, inventory management, and quality control.
- Adaptive Production Planning: AI helps manufacturing industries to optimize production planning and scheduling by leveraging the AI algorithms and optimization techniques. AI-driven manufacturing systems can optimize the scheduling in real time by considering factors such as resource availability, demand forecasting, and production constraints.
- Human-Machine Collaboration: AI-powered systems improve communication between human and machine by providing real-time guidance and decision-making support that ultimately enhance worker productivity. Workers can communicate through natural language interfaces and wearable devices.

- Continuous Improvement and Adaptive Manufacturing: AI and machine learning algorithms can identify process anomalies and suggest the corrective actions that enable adaptive manufacturing to follow the market trends.

3. 3D printing in IM: It plays a significant role in intelligent manufacturing by enabling advanced capabilities and production processes. The following features show the contribution of additive manufacturing in intelligent manufacturing [8].

- Design Flexibility: Conventional manufacturing methods do not provide flexibility in design, but 3D printing allows for design flexibility. It has the capacity to produce intricate geometries, structures, and products that may be impossible using traditional techniques.
- Rapid Prototyping and Iteration: While considering the quality of quick production of physical prototypes directly from digital designs, it speeds up the prototyping. This quick pace enables iterative design improvements that reduce time-to-market.
- Lightweight and Material Efficiency: 3D printing helps to create lightweight and complex structure, which is a requirement of most of the industries such as automotive, aerospace, and healthcare. These industries demand weight reduction that is prominent for performance and fuel efficiency.
- Tooling and Jig Production: Compared to conventional techniques, 3D printing offers faster and more cost-effective solution tooling solutions that ultimately improve the adaptability and agility of intelligent manufacturing systems.

4. AR/VR-assisted IM: These technologies play an important role in the implementation of intelligent manufacturing by enhancing different aspects of manufacturing processes. The following features are as follows [12]:

- Training and Skills Development: Workers can learn the complex processes, assembly tasks, and operate machinery by virtual simulations and interactive training sessions. It improves the worker skills by reducing the learning curve.
- Design Visualization and Collaboration: Designers can examine virtual prototypes with the help of AR/VR devices and suggest the modifications accordingly before the final physical production. It will enhance design accuracy, lessen the design iterations, and improve collaboration efficiency.
- Maintenance and Repair: Workers used AR headsets to overlay digital information onto physical equipment, displaying real-time information for troubleshooting, maintenance procedures, and repair actions. It will increase the efficiency of maintenance operations by reducing equipment downtime.
- Process Optimization and Continuous Improvement: By using AR/VR tools, manufacturers can simulate and analyze the virtual scenarios of production lines, facilities, and workflow processes to identify bottlenecks and optimize workflows.

5. Mobile Robots-assisted Intelligent Manufacturing: Autonomous robots, also known as mobile robots, play a significant role in the implementation of intelligent manufacturing. They increase the efficiency, automation, and flexibility of the system. The following features show the participation of autonomous robots in intelligent manufacturing [13]:

   • Material Handling and Logistics: Mobile robots embedded with sensors and navigation systems autonomously navigate through the manufacturing floor, warehouses, and storage areas to perform different tasks like pick up, transport, and deliver parts to required locations. It will reduce the labor tasks surrounding manual material handling and logistics operations.

   • Assembly and Production: As these robots are equipped with sensors and AI algorithms, they can perform complex assembly and production tasks such as fastening screws, painting, and welding with precision and consistency.

   • Quality Control and Inspection: These robots are embedded with sensors, cameras, and image recognition systems to perform visual inspections, defects detection, and data collection on different product quality parameters.

   • Adaptability and Flexibility: These robots are designed in such a way that they have high adaptability and flexibility in the implementation of intelligent manufacturing. These can be easily reconfigured and reprogrammed and dynamically adjust their routes, tasks, and speed based on real-time data. It will enhance the overall efficiency of the system.

6. Cloud-enabled Intelligent Manufacturing: In this system, the cloud computing technologies and services are integrated into the manufacturing processes. The following features show the contribution of cloud computing in intelligent manufacturing [14]:

   • Scalable Computing Resources: Cloud computing handles a huge amount of data generated by sensors, machines, and other sources by providing scalable computing resources such as network bandwidth, storage capacity, and processing power. Manufacturers can easily adjust their computing resources based on their requirements.

   • Data Storage and Analytics: Cloud computing leverages machine learning and AI algorithms to store and analyze huge amounts of data. This data may be related to manufacturing, production, supply chain, machine performance, and quality control.

   • Remote Monitoring and Management: The data generated from the connected devices transfers to the cloud permitting the managers and operators to monitor the system remotely. It will enable proactive maintenance, timely interventions, and efficient resource management.

   • On-demand Software as a Service (SaaS): Manufacturers can get the special tools and applications without the need of local installations with the help of cloud-based SaaS solutions. It offers multiple tools to manufacturers such as ERP systems, quality management systems, MES, and supply chain management solutions.

7. Cybersecure Intelligent Manufacturing: While performing the tasks remotely there is always a threat of cyberattacks to the system. Cybersecure intelligent manufacturing refers to the integration of efficient and robust cybersecurity measures in the implementation of intelligent manufacturing. It protects the manufacturing system from unauthorized access. The following features will demonstrate the corporation of cybersecure intelligent manufacturing [15].

- Threat Detection and Prevention: It includes the implementation of proactive threat detection and prevention systems having firewalls, intrusion detection systems, and security information and event management systems. This will lead to monitoring the network traffic and identifying the potential threats to protect the system while implementing intelligent manufacturing.
- Secure Network Infrastructure: It involves the establishment of secure network infrastructure by proposing efficient and secure protocols to protect the system under consideration.
- Data Protection and Privacy: Strong encryption algorithms are used to protect the data from unauthorized access. These algorithms may be symmetric and asymmetric. Symmetric algorithms use only one key for encryption and decryption purposes, unlike asymmetric algorithms which use two keys.
- Regulatory Compliance: It includes the identification of potential vulnerabilities in intelligent manufacturing implementation systems by including regular security assessments, penetrating testing, and vulnerability scans. Compliance with regulations, including ISO 27001 or NIST Cybersecurity framework ensures effective cybersecurity practices and controls.

D-P plane: This plane is related to smart factory for Personalized Customization. It focuses on the efficient production of highly customized products by leveraging advanced technologies and driven systems to meet customer requirements. It contains the following stages [15]:

Customer Data Integration: With the aim to get the customer input and design specifications, this factory offers different sources such as configurators and online platforms, CRM systems, and social media. It will help customers customize their products features. By analyzing this data, it is used to generate digital product models.

Product Configuration: The smart factory offers the product configuration based on individual needs after the collection of customer data. Customization options are guided by firm's production capabilities and constraints, customers can customize their required products by choosing different options like size, color, feature, and material.

Digital Design and Simulation: The smart factory offers advanced simulation tools like CAD to convert the customer's configurations into digital models to prevent any conflict or potential issues related to design early in the process. These tools are also helpful for virtual testing and visualization of the customized products to keep ensure feasibility and manufacturability.

Real-time Data Exchange: The smart factory integrates and exchanges data from different sources like production systems, customer orders, and supply chain partners to facilitate seamless communication throughout the manufacturing ecosystem. Manufacturers can get different information like production status, resource allocation, and material availability by connecting various stages of production.

Customer Engagement, Traceability, and Feedback: The smart factory allows manufacturers to engage customers by offering delivery updates and personalized notifications. This factory also offers technologies like barcode and RFID tags to trace the customized products throughout the manufacturing process to enable customers with real-time updates and transparency. This interactive interface for customization helps to enhance customer feedback.

2.2 PHYSICAL INTEGRATION IN CYBER PHYSICAL SPACE

2.2.1 OSI – Open System Interconnection Standard

The Open Systems Interconnect (OSI) standard is a conceptual framework that defines a set of protocols and guidelines for designing and implementing network communication systems. Developed by the International Organization for Standardization (ISO) in the late 1970s, the OSI model provides a structured approach to network architecture and promotes interoperability between different vendors' networking technologies. In this section, we will explore the OSI standard, including its seven layers, their functions, and the impact of the OSI model on network communication. The OSI model consists of seven layers, each responsible for specific functions in the network communication process. These layers are [16]:

Physical Layer: The physical layer is responsible for the actual transmission of raw data over a physical medium, such as copper cables, fiber optics, or wireless channels. It defines the electrical, mechanical, and procedural characteristics required for physical transmission. This layer deals with issues such as voltage levels, signal timing, and data encoding.

Data Link Layer: The data link layer provides reliable transmission of data frames between adjacent network nodes. It detects and corrects errors that may occur at the physical layer and ensures that data is transmitted without errors or loss. This layer is responsible for framing, error detection, flow control, and access control.

Network Layer: The network layer enables the routing of data packets across multiple networks. It determines the best path for data transmission based on the network topology, addressing, and congestion control. The network layer defines logical addressing and routing protocols to ensure the efficient and reliable delivery of data.

Transport Layer: The transport layer ensures end-to-end reliable delivery of data between hosts. It segments and reassembles data into smaller units, known as segments, and provides mechanisms for error recovery, flow control, and congestion control. This layer establishes connections, maintains data sequencing, and manages the quality of service.

Session Layer: The session layer establishes, maintains, and terminates sessions between applications running on different network devices. It provides synchronization points in data exchange, allowing for the coordination of multiple communication streams. The session layer also manages authentication, authorization, and session recovery.

Presentation Layer: The presentation layer is responsible for the translation, compression, and encryption of data. It formats the data to be presented in a common representation that can be understood by the application layer. This layer handles tasks such as data compression, encryption, and data format conversion.

Application Layer: The application layer interacts directly with end-user applications. It provides a means for users to access network services and supports application-specific protocols, such as HTTP for web browsing or SMTP for email communication. The application layer is responsible for managing user interfaces, file transfers, and network services.

The OSI model's layering approach allows for the separation of different functions and promotes modular design. Each layer performs a specific set of tasks, and the interfaces between layers define how they communicate with each other. This modular design enables flexibility, scalability, and interoperability, as different layers can be implemented independently and replaced or upgraded without affecting other layers.

One of the key advantages of the OSI model is its promotion of interoperability between different vendors' networking technologies. By defining a standardized set of protocols and interfaces, the OSI model allows devices from different manufacturers to communicate seamlessly. This interoperability has played a crucial role in the growth and development of computer networks, enabling the creation of heterogeneous networks with diverse hardware and software components.

The OSI model has also influenced the development of actual networking protocols and technologies. While the OSI protocols themselves are not widely used in practice, the conceptual framework provided by the OSI model has been instrumental in the development of other networking protocols, such as the Transmission Control Protocol/Internet Protocol (TCP/IP), which is the foundation of the modern internet.

TCP/IP, which is widely used in today's networks, is based on a similar layering concept as the OSI model. However, TCP/IP combines multiple layers of the OSI model into fewer layers, resulting in a simpler and more practical implementation. The TCP/IP model consists of four layers: Network Interface Layer, Internet Layer, Transport Layer, and Application Layer. Despite the differences in the number of layers and specific protocols used, the TCP/IP model and the OSI model share the common goal of facilitating network communication and interoperability.

In a nutshell, the OSI standard provides a structured approach to network architecture and defines a set of protocols and guidelines for designing and implementing network communication systems. The OSI model consists of seven layers, each responsible for specific functions in the network communication process. It promotes interoperability between different vendors' networking technologies and has influenced the development of actual networking protocols, such as TCP/IP. While the OSI protocols themselves are not widely used, the OSI model's conceptual framework has had a significant impact on the field of network communication, enabling the growth and development of computer networks.

2.3 LEGACY SERIAL AND PARALLEL COMMUNICATION PROTOCOLS

Serial and parallel communication protocols have been fundamental in the field of data transmission for several decades. These protocols allowed computers and other devices to exchange information, enabling the development of early networking technologies and facilitating the transfer of data between different components. In this section, we will explore legacy serial and parallel communication protocols, their characteristics, advantages, and limitations.

2.3.1 SERIAL COMMUNICATION PROTOCOLS

Serial communication protocols transmit data sequentially, one bit at a time, over a single data line. This method of communication is widely used in various applications, ranging from simple device connections to complex network systems. Here are a few examples of legacy serial communication protocols [17]:

2.3.2 RS-232 (RECOMMENDED STANDARD 232)

RS-232, also known as the serial port or COM port, was one of the earliest and most widely used serial communication protocols. It defined the electrical and mechanical characteristics for connecting devices, such as modems, printers, and early personal computers. RS-232 used a voltage-level signaling system with positive and negative voltages to represent binary data. It supported full-duplex communication, allowing simultaneous data transmission and reception. RS-232 had a maximum data rate of 20 kbps (kilobits per second) over short distances, but its performance degraded with longer cable lengths. It used a relatively large number of signal lines, including data lines, control lines, and ground, which required multiple pins on the connector. Despite its limitations, RS-232 played a vital role in the early days of computer communication and remains in use today for certain applications.

2.3.3 RS-485

RS-485 is a serial communication protocol that addresses the limitations of RS-232 for longer distances and multi-node communication. It is designed for use in industrial automation, process control systems, and other applications that require reliable long-distance communication. RS-485 uses differential signaling, where the binary data is represented by the voltage difference between two data lines, providing better noise immunity and enabling longer cable runs.

Unlike RS-232, RS-485 supports multi-drop communication, allowing multiple devices to be connected to a single data line. It uses a master-slave architecture, where the master device controls the communication by enabling or disabling the transmission from individual slave devices. RS-485 offers higher data rates than RS-232, typically ranging from 100 kbps to 10 Mbps, depending on cable length and other factors.

2.3.4 PARALLEL COMMUNICATION PROTOCOLS

Parallel communication protocols transmit multiple bits simultaneously over multiple data lines. This method was commonly used in early computer systems and peripheral devices to achieve high-speed data transfer. A couple of examples of legacy parallel communication protocols follow [18].

2.3.5 CENTRONICS PARALLEL INTERFACE

The Centronics parallel interface, commonly known as the printer port, was widely used for connecting printers to computers in the past. It used a parallel data transmission method with a fixed number of data lines (typically 8 or 16) to send data in parallel, resulting in faster data transfer compared to serial protocols. The Centronics interface was bidirectional, supporting both data transmission and reception. The printer port operated in a synchronous manner, with a handshake mechanism to control the flow of data between the computer and the printer. However, the data transfer speed of the Centronics interface was limited to a maximum of a few megabits per second, and its use declined with the introduction of more advanced printer connectivity options, such as USB.

2.3.6 SCSI (SMALL COMPUTER SYSTEM INTERFACE)

SCSI is a parallel interface primarily used for connecting high-performance peripheral devices, such as hard drives, scanners, and optical drives, to computer systems. It provided fast data transfer rates and supported multiple devices on the same bus. SCSI was highly flexible, allowing for daisy-chaining multiple devices and offering a wide range of commands for device control and data transfer. SCSI interfaces used a combination of data lines, control lines, and termination to ensure reliable communication. Various SCSI standards were developed over the years, including SCSI-1, SCSI-2, and SCSI-3, each introducing improvements in data transfer rates and capabilities. Despite its advantages, SCSI was relatively expensive to implement and required specialized hardware, limiting its widespread adoption.

2.3.7 ADVANTAGES AND LIMITATIONS

Serial and parallel communication protocols offered distinct advantages and faced specific limitations.

2.3.8 ADVANTAGES OF SERIAL COMMUNICATION

Simplicity: Serial protocols required fewer data lines, making them easier to implement and reducing the complexity of connectors and cables.

Longer Distances: Serial communication was suitable for longer cable runs, as it suffered less from signal degradation compared to parallel protocols.

Multi-drop Capability: Some serial protocols, such as RS-485, allowed multiple devices to share a single data line, reducing the number of connections required.

2.3.9 LIMITATIONS OF SERIAL COMMUNICATION

Lower Data Rates: Serial protocols generally offered slower data transfer rates compared to parallel protocols, making them less suitable for applications requiring high-speed data transmission.

Limited Bidirectional Communication: Serial protocols often supported half-duplex communication, where data could be transmitted in only one direction at a time, which could introduce delays in bidirectional communication.

Complexity in Multi-node Communication: Achieving multi-node communication in serial protocols required additional control mechanisms, increasing complexity and potentially reducing performance.

2.3.9.1 Advantages of Parallel Communication

Higher Data Rates: Parallel protocols enabled faster data transfer rates by transmitting multiple bits simultaneously, making them suitable for applications that required high-speed data transmission.

Simultaneous Data Transfer: Parallel protocols allowed for simultaneous data transfer between devices, potentially reducing latency and improving overall system performance.

2.3.9.2 Limitations of Parallel Communication

Increased Complexity: Parallel protocols required a larger number of data lines, resulting in more complex connectors and cables.

Signal Integrity Challenges: Maintaining signal integrity over multiple data lines became increasingly difficult with higher data rates and longer cable lengths, leading to potential signal degradation and data errors.

Cost: Implementing parallel protocols involved additional hardware complexity and cost, making them less suitable for cost-sensitive applications.

In a nutshell, legacy serial and parallel communication protocols played a crucial role in the early days of data transmission. Serial protocols, such as RS-232 and RS-485, provided reliable communication over shorter and longer distances, respectively. Parallel protocols, such as Centronics and SCSI, facilitated high-speed data transfer between devices. Each type of protocol had its own advantages and limitations, and their use depended on the specific requirements of the application. However, with advancements in technology, modern protocols have largely replaced legacy serial and parallel protocols, offering improved perform-ance, scalability, and ease of use.

2.4 MODERN SERIAL AND PARALLEL COMMUNICATION PROTOCOLS

In the rapidly evolving world of technology, serial and parallel communication protocols continue to play a vital role in data transmission. However, modern protocols have emerged to address the limitations of legacy protocols and meet the increasing demands of high-speed, reliable, and efficient communica-tion. In this section, we will explore some of the modern serial and parallel communication protocols, their characteristics, advantages, and applications [18].

2.4.1 MODERN SERIAL COMMUNICATION PROTOCOLS

2.4.1.1 Universal Serial Bus (USB)

The Universal Serial Bus (USB) protocol has become ubiquitous in modern computing systems, offering a versatile and standardized interface for connecting a wide range of devices. USB supports hot-swapping, meaning devices can be connected and disconnected without rebooting the system. It provides both power and data transfer capabilities over a single cable, simplifying connectivity. USB offers several versions, including USB 1.1, USB 2.0, USB 3.0, USB 3.1, and the latest USB 3.2 and USB4 standards. Each version introduces improvements in data transfer rates, power delivery, and features. USB 3.0 and later versions employ a serial data transfer method with multiple lanes, allowing for faster data rates. USB 3.0 offers data rates up to 5 Gbps, USB 3.1 Gen 2 supports up to 10 Gbps, and USB 3.2 and USB4 can achieve up to 20 Gbps or higher. USB is widely used for various devices, including keyboards, mice, external storage devices, printers, smartphones, and more. Its popularity is attributed to its ease of use, compatibility across different platforms, and the availability of a wide range of peripherals.

2.4.1.2 Thunderbolt

Thunderbolt is a high-speed serial communication protocol developed by Intel in collaboration with Apple. It combines PCIe (Peripheral Component Interconnect Express) and DisplayPort protocols, enabling data transfer and video display capabilities over a single cable. Thunderbolt interfaces provide high data transfer rates, making it suitable for tasks like video editing, high-resolution displays, and fast storage devices.

Thunderbolt 3 is the most recent version, offering data transfer rates of up to 40 Gbps. It supports daisy-chaining of devices, meaning multiple Thunderbolt devices can be connected in a series, simplifying cable management. Thunderbolt 3 ports also support USB-C connectors, making them versatile for use with USB-C devices. Thunderbolt technology has found its place in professional environments where high-speed data transfer and low latency are critical, such as video production, audio interfaces, and external storage systems.

2.4.2   MODERN PARALLEL COMMUNICATION PROTOCOLS

2.4.2.1   Peripheral Component Interconnect Express (PCIe)

PCIe is a high-speed parallel communication protocol commonly used to connect various internal computer components, such as expansion cards, graphics cards, network adapters, and storage devices. PCIe provides a point-to-point serial communication architecture, allowing each device to have dedicated communication channels [19].

PCIe offers multiple generations, including PCIe 1.0, PCIe 2.0, PCIe 3.0, PCIe 4.0, and PCIe 5.0. Each generation provides improvements in data transfer rates, bandwidth, and power management. PCIe 4.0, for instance, offers a data transfer rate of 16 Gbps per lane, while PCIe 5.0 doubles it to 32 Gbps per lane.

One significant advantage of PCIe is its scalability. PCIe devices can be connected through different slot sizes, including x1, x4, x8, and x16, allowing for flexible configurations based on specific requirements. The scalability and high data transfer rates make PCIe a preferred choice for high-performance computing applications, such as gaming PCs, workstations, and servers.

2.4.2.2   Serial Attached SCSI (SAS)

SAS is a high-speed, point-to-point serial communication protocol designed for connecting storage devices, such as hard disk drives (HDDs), solid-state drives (SSDs), and tape drives. SAS builds upon the SCSI protocol, offering faster data transfer rates and enhanced performance compared to parallel SCSI.

SAS provides full-duplex communication and supports multiple devices on a single bus. It offers higher data transfer rates compared to its predecessor, Parallel SCSI, and enables the connection of a larger number of devices in a more efficient manner. SAS can achieve data transfer rates up to 24 Gbps per lane, with multiple lanes supporting even higher aggregate speeds.

SAS is commonly used in enterprise-level storage systems and data centers, where speed, reliability, and scalability are essential. It allows for the simultaneous access of multiple devices, making it suitable for demanding applications that require fast and efficient data retrieval.

2.4.3   ADVANTAGES AND APPLICATIONS

Modern serial and parallel communication protocols offer several advantages over their legacy counterparts.

2.4.3.1   Advantages of Modern Serial Communication Protocols

Higher Data Transfer Rates: Modern serial protocols provide significantly faster data rates, allowing for efficient transmission of large volumes of data within shorter time frames.

Simplicity and Ease of Use: Modern serial protocols often have simpler cabling requirements and support plug-and-play functionality, making them user-friendly and convenient to deploy.

Enhanced Compatibility: Modern serial protocols, such as USB, are designed to be backward compatible, ensuring compatibility with legacy devices while offering improved capabilities.

Higher Bandwidth: Modern parallel protocols, like PCIe, offer increased bandwidth and faster data transfer rates, making them suitable for high-performance applications that require the efficient handling of large data streams.

Scalability: Parallel protocols like PCIe allow for scalable configurations, accommodating a wide range of devices and expanding system capabilities as needed.

Low Latency: Parallel communication protocols minimize latency due to their ability to transfer multiple bits simultaneously, allowing for real-time and high-performance applications.

2.4.3.2 Applications of Modern Serial and Parallel Communication Protocols

High-speed Storage Systems: Modern serial and parallel protocols, such as Thunderbolt, PCIe, and SAS, are commonly used in storage systems to provide fast and reliable data transfer rates between storage devices and host systems.

Multimedia and Video Production: Serial protocols like Thunderbolt and USB are employed for connecting high-resolution displays, cameras, audio interfaces, and other multimedia devices, allowing for efficient data transfer and real-time streaming.

Networking and Communication: Serial protocols like USB and Ethernet are extensively used in networking equipment, routers, switches, and network adapters to facilitate reliable and high-speed communication between devices.

High-performance Computing: Parallel protocols like PCIe find applications in high-performance computing systems, gaming PCs, and workstations, where fast data transfer rates and low latency are crucial for processing-intensive tasks.

In a nutshell, modern serial and parallel communication protocols have evolved to meet the increasing demands of high-speed, reliable, and efficient data transmission. Serial protocols like USB and Thunderbolt offer versatile connectivity options for a wide range of devices, while parallel protocols like PCIe and SAS provide high bandwidth and low latency for demanding applications. These protocols have transformed the landscape of data transmission, enabling seamless communication and supporting advancements in various fields, including storage systems, multimedia, networking, and high-performance computing.

2.5   CONCLUSION

The theme of the chapter is to implement facilities provided by the SME policy to sectoral-free SME or startup to flare up the economy by including less-trained workers to have more training (Paid or Unpaid), start a business to that of a gig economy level, elevate themselves to a startup level, and converge to SME in 3–5 years. This proposition is especially useful to young generations due to recession and socio-economic political factors; this happens to be the only way out. It is so that most families are applying this scheme of work without any training and knowledge of related local and international laws. Facilitation by relevant NGOs and government bodies will bring in the desired change of a stimulated economy.

REFERENCES

[1] Zhong, R. Y., et al. "Intelligent manufacturing in the context of industry 4.0: a review." *Engineering* 3.5 (2017): 616–630.

[2] Liang, S., et al. "Intelligent manufacturing systems: a review." *International Journal of Mechanical Engineering and Robotics Research* 7.3 (2018): 324–330.

[3] Khan, W. A. and SI, A. R. *Standards for engineering design and manufacturing.* CRC Press (2005).

[4] Singh, M., Goyat, R., and Panwar, R. "Fundamental pillars for industry 4.0 development: implementation framework and challenges in manufacturing environment." *The TQM Journal* 36 (2023):288–309.

[5] Liu, L., Yan, C.-B., and Li, J. "Modeling, analysis, and improvement of batch-discrete manufacturing systems: A systems approach." *IEEE Transactions on Automation Science and Engineering* 19.3 (2021): 1567–1585.

[6] Srai, J. S., et al. "Future supply chains enabled by continuous processing—Opportunities and challenges. May 20–21, 2014 Continuous Manufacturing Symposium." *Journal of Pharmaceutical Sciences* 104.3 (2015): 840–849.

[7] Singh, R. *Introduction to basic manufacturing processes and workshop technology.* New Age International (2006).

[8] Huang, Q. "Intelligent manufacturing." *Understanding China's manufacturing industry.* Springer Nature Singapore (2022): 111–127.

[9] Zhong, R. Y. and Ge, W. "Internet of Things enabled manufacturing: a review." *International Journal of Agile Systems and Management* 11.2 (2018): 126–154.

[10] Samanta, H., et al. "IoT-based intelligent manufacturing system: A review." *Intelligent manufacturing management systems: Operational applications of evolutionary digital technologies in mechanical and industrial engineering.* Wiley (2023): 59–84.

[11] Wan, J., et al. "Artificial-intelligence-driven customized manufacturing factory: key technologies, applications, and challenges." *Proceedings of the IEEE* 109.4 (2020): 377–398.

[12] Eswaran, M. and Bahubalendruni, M. V. A. R. "Challenges and opportunities on AR/VR technologies for manufacturing systems in the context of industry 4.0: A state of the art review." *Journal of Manufacturing Systems* 65 (2022): 260–278.

[13] Liu, X. J. "Research toward IoT and robotics in intelligent manufacturing: a survey." *International Journal of Materials, Mechanics and Manufacturing* 7.3 (2019): 128–132.

[14] Tang, H., et al. "A reconfigurable method for intelligent manufacturing based on industrial cloud and edge intelligence." *IEEE Internet of Things Journal* 7.5 (2019): 4248–4259.

[15] Zhang, X., Ming, X., and Bao, Y. "A flexible smart manufacturing system in mass personalization manufacturing model based on multi-module-platform, multi-virtual-unit, and multi-production-line." *Computers & Industrial Engineering* 171 (2022): 108379.

[16] Aschenbrenner, J. R. "Open systems interconnection." *IBM Systems Journal*, 25.3.4 (1986): 369–379.

[17] Dawoud, D. S. and Dawoud, P. *Serial communication protocols and standards.* CRC Press (2020).

[18] Anitha, T., et al. "A Review on Communication Protocols of Industrial Internet of Things." in *2022 2nd International Conference on Computing and Information Technology (ICCIT)*, (Jan. 2022), pp. 418–423.

[19] Chakravarthi, V. S. "M2M communication protocols." *Internet of Things and M2M communication technologies, architecture and practical design approach to IoT in Industry 4.0.* Springer (2021): 167–190.

Part 2

Modern Control Theory

3 Integrating Fault Tolerance in Feedback Control Systems

Ali Nasir

3.1 INTRODUCTION

Traditional feedback control systems rely heavily on the proper functionality of actuators and sensors. Other than that, most feedback control systems assume a specific dynamical model of the system to be controlled (often called the "plant"). Any deviation in the dynamical model of the plant or a fault in an actuator or a sensor can affect the overall performance of a feedback control system. Therefore, it is very important to integrate fault tolerance in feedback control systems so that effect of a fault in a sensor, an actuator, or the plant may be minimized on the overall performance of the feedback control system.

Fault-tolerant control systems have been a focus of attention for multiple decades. A comprehensive review of different methods used for integrating fault tolerance in feedback control systems has been presented in [1]. A holistic view of the three main components of a fault-tolerant control system is discussed in [2]. These components include a fault detection mechanism, a fault diagnosis (or identification) scheme, and a control reconfiguration capability. Similar discussions of the structure and implications of fault-tolerant control are available in [3] and [4]. As mentioned earlier, the purpose of the design of fault-tolerant control is to minimize the degradation in the performance of the feedback control. In this regard, a unique method that explicitly considers the control system performance in the design of fault-tolerant control is discussed in [5].

One of the most vulnerable equipment in terms of being prone to faults in feedback control systems is the sensor. Sensor faults have been discussed generally in [6] and specifically for spacecraft missions in [7]. Traditionally, feedback control systems rely on state estimators in addition to the sensors. This fact has been utilized for constructing state estimator-based fault diagnosis schemes such as integrated multiple-model-based state estimation approaches [8,9]. In these approaches, post-fault models are constructed beforehand and the actual behavior of the system is compared with these models in order to identify the exact nature of the fault.

DOI: 10.1201/9781003376620-5

Multiple books have been published on the topic of fault-tolerant control [10,11]. A recent review on the advancements on the fault tolerance in feedback control systems is presented in [12]. Latest applications of fault-tolerant control include robotic manipulators [13], marine surface vessels [14], and spacecraft missions [15].

The purpose of this chapter is to introduce fault-tolerant control for the students who are new to the field but have some preliminary knowledge regarding feedback control systems. This chapter begins with some basic properties of feedback control systems that are important for understanding the integration of fault tolerance with feedback control systems. Afterwards, the main modules of the fault-tolerant control are discussed with appropriate examples. Future directions in fault-tolerant control are provided toward the end of the chapter.

3.2 PRELIMINARIES OF FEEDBACK CONTROL

This section discusses the basic properties of feedback control systems that are important in the design of fault-tolerant control. All of the discussions in this chapter are with respect to the linear time-invariant systems (because these systems are easy to understand). However, the concepts discussed here can be generalized to the nonlinear and time-varying systems at least under specific conditions, e.g., small signal variation.

Consider the following linear time-invariant system:

$$\begin{aligned} \dot{x} &= Ax + Bu \\ y &= Cx \\ x &\in \mathbb{R}^n, u \in \mathbb{R}^m, y \in \mathbb{R}^p \end{aligned} \qquad (3.1)$$

There are three key properties of this system that every control systems engineer should know about, i.e., stability, controllability, and observability. The stability property is relevant to the A matrix in the system model. Specifically, the system (open loop) is deemed as stable (in the sense of Lyapunov) if all of the eigen values of the A matrix have negative real part with no repeated eigen values with zero real part. Furthermore, if all of the eigen values of the A matrix have strictly negative real part, then the system (open loop) is asymptotically stable. For closed-loop stability, with control design of the form $u = -Kx$, the eigen values of the matrix $(A - BK)$ are used to determine the stability. Practically, the stability means that we are able to keep the system in certain condition (referred to as the equilibrium point condition) despite small disturbances.

Next important property of a feedback control system is controllability. Controllability is the ability to drive the states of the system from any arbitrary initial condition to any arbitrary final value with the help of a piecewise continuous finite valued control signal in finite time. Mathematically, the system is said to be controllable if and only if the matrix $Q = [B \quad AB \quad A^2B \quad \dots \quad A^{n-1}B]$ has rank n. Practically, the controllability means that we have enough actuators in the system that are able to maneuver it from any point to any other point in finite time and using finite energy.

Finally, the observability property of the system is the ability to determine the initial state of the system by using information about the control signal(s) u and the output signal(s) y. Basically, the ability to determine the initial condition gives us the ability to determine the state value x at any time by solving the differential equation governing the system behavior. Mathematically, the system is said to be observable if and only if the matrix $O = [C \quad CA \quad CA^2 \quad ... \quad CA^{n-1}]^T$ has rank n. Practically, the observability means that we have enough sensors in the systems that we are able to calculate the entire system state using the sensor values.

Example 1: Consider the following system:

$$\dot{x} = \begin{bmatrix} 0 & 1 \\ 0 & 0 \end{bmatrix} x + \begin{bmatrix} 0 \\ 1 \end{bmatrix} u, \quad x = \begin{bmatrix} x_1 \\ x_2 \end{bmatrix} \in \mathbb{R}^2$$

$$y = [1 \quad 0] x$$

i. Determine if the system is observable or not.
ii. Change y to $y = [0 \quad 1] x$ and recalculate observability.

Solution

i. The observability matrix for the system is given by:

$$O = \begin{bmatrix} C \\ CA \end{bmatrix} = \begin{bmatrix} 1 & 0 \\ 0 & 1 \end{bmatrix}$$

Since both columns of the O are linearly independent, the rank of O matrix is 2 and hence the system is observable (because n is also equal to 2).

ii. The observability matrix for new output is given by:

$$O = \begin{bmatrix} C \\ CA \end{bmatrix} = \begin{bmatrix} 0 & 1 \\ 0 & 0 \end{bmatrix}$$

Note that one of the columns of O is zero. Hence the two columns are not linearly independent. The rank of O in this case is 1 which is less than n. Therefore, the system is not observable.

Example 2: Determine the controllability property of the system in example 1.
Solution
The controllability matrix for this system is given by:

$$Q = [B \quad AB] = \begin{bmatrix} 0 & 1 \\ 1 & 0 \end{bmatrix}$$

Since both columns of the Q matrix are linearly independent[1], the rank of Q matrix is 2. Also, the size of the system state vector is also 2 ($n = 2$). Therefore, the system is controllable.

Note that a square matrix is full rank if and only if its determinant is nonzero. Since Q is a square matrix in this example, we could also use the determinant test in order to determine the controllability.

Example 3: Consider the following system:

$$\dot{x} = \begin{bmatrix} 0 & 1 \\ 0 & 0 \end{bmatrix} x + \begin{bmatrix} 0 \\ 1 \end{bmatrix} u, \; x = \begin{bmatrix} x_1 \\ x_2 \end{bmatrix} \in \mathbb{R}^2$$

i. Determine the open-loop stability of the system.
ii. Consider $u = -x_1 - 2x_2$, determine the closed-loop stability of the system.

Solution

i. The open-loop stability is determined by the eigen values of the matrix $\begin{bmatrix} 0 & 1 \\ 0 & 0 \end{bmatrix}$. Since the matrix is upper triangular, its eigen values are the same as the diagonal elements, i.e., {0,0}. Since there are repeated eigen values with zero real part, the system is open-loop unstable.
ii. The given control law can be rewritten as $u = -Kx = -[1 \; 2]x$. Therefore, the closed-loop stability is determined using the eigen values of the following matrix:

$$\begin{bmatrix} 0 & 1 \\ 0 & 0 \end{bmatrix} - \begin{bmatrix} 0 \\ 1 \end{bmatrix} [1 \; 2] = \begin{bmatrix} 0 & 1 \\ -1 & -2 \end{bmatrix}$$

The eigen values are calculated as:

$$\det\left(\begin{bmatrix} \lambda & 0 \\ 0 & \lambda \end{bmatrix} - \begin{bmatrix} 0 & 1 \\ -1 & -2 \end{bmatrix}\right) = 0 \rightarrow \lambda = \{-1, -1\}$$

Since both of the eigen values have strictly negative real parts, the closed-loop system is asymptotically stable.

3.3 BASIC ARCHITECTURE OF FAULT-TOLERANT CONTROL

A general architecture of fault-tolerant control is shown in Figure 3.1. Here, u is the control input vector, y is the output vector, $\hat{x}$ is the estimate of the state

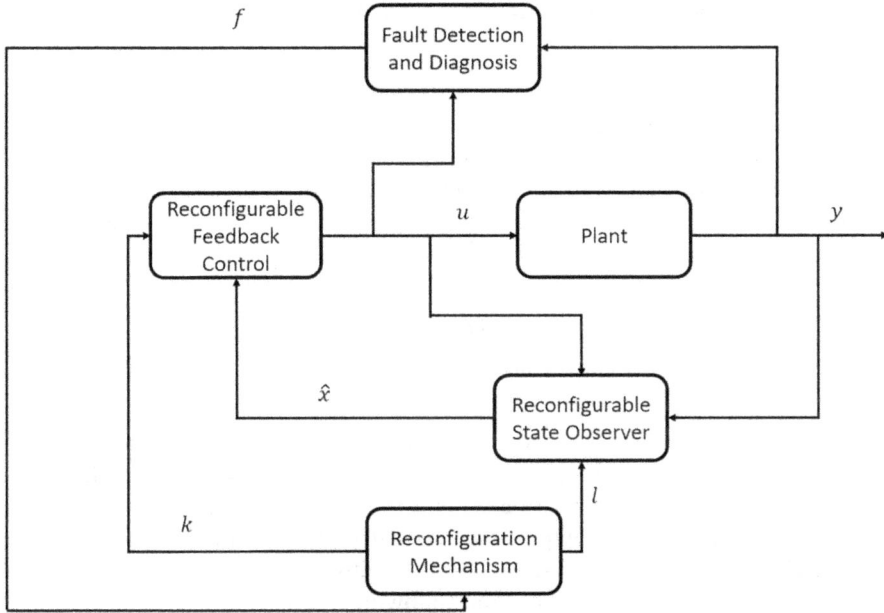

FIGURE 3.1 General architecture of fault-tolerant control.

vector, f is the fault vector with binary flags indicating which fault has occurred and which fault has not occurred, k is variable that guides the reconfigurable feedback control about the selection of appropriate control scheme among the available alternatives, l is variable that guides the reconfigurable state observer about the selection of appropriate state estimation scheme among the available alternatives. During the closed-loop control operation, the information of the input and the output is shared with fault detection and diagnosis module which compares the actual values of the input and output with their expected values. If there is any discrepancy, built-in diagnosis methods are used for identification of the fault and generation of the appropriate fault vector that is communicated to the reconfiguration mechanism. The reconfiguration mechanism (upon receiving the fault vector) uses its own built-in methods to determine the best available control scheme and the best available state estimation scheme that minimizes the degradation in the post-fault performance of the control system. The information regarding the best post-fault control scheme is then sent to the reconfigurable feedback control module. Similarly, the information regarding the best post-fault state estimation scheme is sent to the reconfigurable state estimation module. In this manner, the performance of the feedback control system is managed despite the occurrence of faults. Note that the architecture that has been presented here assumes prior knowledge regarding the post-fault behavior of the system and regarding the types of faults that could occur.

Next, we investigate the details of fault detection and diagnosis as well as the reconfiguration mechanism.

3.4   FAULT DETECTION AND DIAGNOSIS

Fault detection and diagnosis involves two tasks. First is the detection of an occurrence of a fault. Second is the identification of the fault. Fault detection is usually based on thresholds. For example, suppose that a feedback control system is initialized with output $y(0) = y_0$ at time $t = 0$. The control will maneuver the output according to the control objectives. Meanwhile the fault detection uses the information of the control signal and the output in order to predict the values of the output using the equations:

$$\bar{x}(t) = \int_0^t [A\bar{x}(\tau) + Bu(\tau)]d\tau + \bar{x}_0$$
$$\bar{y}(t) = C\bar{x}(t) \tag{3.2}$$

Here, $\bar{x}(t)$ is the simulated value of the state vector at time t and $\bar{y}(t)$ is the simulated value of the output vector at time t. Also, $\bar{x}_0 = Cy(0)$ is the initial state for the simulation which is the same as the initial state in the actual system. The simulated value of the output is compared with the actual output and if the difference between the magnitude of the two output vectors exceeds a threshold, then a fault is declared as shown in the following relation:

$$\|y(t) - \bar{y}(t)\| > \eta \rightarrow \textit{fault detected} \tag{3.3}$$

There is a lot of scientific research around the design of the threshold η because if η is too small, then even small system disturbances shall be considered as faults (this situation is referred to as a false alarm). On the other hand, if η is too large, then even at the occurrence of a fault, the difference between the actual and the simulated output may not exceed the value of η (this situation is referred to as a missed detection). In some cases, the threshold comparison itself is time-based, i.e., fault call is generated only if the threshold is consistently violated for a pre-determined duration of time. Such an approach protects against the false alarms in case of a temporary disturbance.

Consequently, there are many ideas related to the design of η. For example, making η variable with respect to the value of the output, e.g., larger threshold for large values of y and smaller threshold for small values of y or specific values of the threshold for different ranges in the values of y. For time-varying systems, η could depend upon time as well.

Once the fault is detected, the next phase is the identification of the fault, i.e., exact nature and cause of the fault. In this regard, a common approach is to develop mathematical models for the anticipated possible faults and compare the actual behavior with all of the faulty-model behaviors. Such an approach is called a multiple-model-based approach. Here again, the design of thresholds is a major challenge. Another challenge is the discrimination among similar faults.

In the context of the system in equation (3.1), there are three types of effects that a fault may have on the system model. A fault within the plant (system to be controlled) shall affect the entries of the matrix A, a fault in an actuator shall change the matrix B, and any fault in a sensor shall cause changes in the matrix C.

It is usually very difficult to deal with simultaneous faults as the number of fault scenarios (and hence the number of fault models) increases exponentially as we begin to incorporate simultaneous faults. For example, consider a feedback control system with two sensors and two actuators. If we consider failure of each individual component separately, then we need four fault models. However, if we also consider simultaneous faults, say two at a time, then we need additional six fault models (one for each possible pair) making a total of ten fault models. Similarly, the of k number of components, the number of additional fault models required shall be $0.5k\,(k-1)$ if pair-wise faults are to be considered. More fault models shall be needed if three at a time or more simultaneous faults are to be considered.

Mathematical depiction of the multiple-model-based fault identification is presented in (4) and (5). Note that there are q fault models where each model has its own unique A, B, and C matrices. All of these models are simulated and after the detection of a fault, the actual output is compared with the simulated output of each of the models. A major difference here in terms of the comparison is that as opposed to the fault detection where we are looking at the difference between the two outputs exceeding a threshold, in fault identification, we are looking for the difference being within a threshold. This is because, in identification, we are interested in the *similarity* of the actual behavior with any of the pre-determined fault models.

Here again, like fault detection, the diagnosis threshold design is a challenging task. An additional challenge here is the possibility of simultaneous multiple detections. Note that such a situation cannot be considered as a multiple faults' scenario because for multiple faults, a separate model is required. For example, if the value of the output matches simultaneously with simulated fault output for fault 1 and fault 2, we would not conclude both faults. Rather a conflict resolution scheme would be required for exact identification of the actual fault. In order to be able to identify both fault 1 and fault 2 simultaneously, we need a separate model with A, B, and C matrices corresponding to the fault situation.

$$\dot{\bar{x}}_1 = A_1\bar{x}_1 + B_1 u$$
$$\bar{y}_1 = C_1\bar{x}_1$$

$$\dot{\bar{x}}_2 = A_2\bar{x}_2 + B_2 u$$
$$\bar{y}_2 = C_2\bar{x}_2 \qquad\qquad (3.4)$$

$$\vdots$$

$$\dot{\bar{x}}_q = A_q\bar{x}_q + B_q u$$
$$\bar{y}_q = C_q\bar{x}_q$$

$$\|y(t) - \bar{y}_1(t)\| < \eta_1 \rightarrow fault_1\ identified$$
$$\|y(t) - \bar{y}_2(t)\| < \eta_2 \rightarrow fault_2\ identified$$
$$\vdots \qquad\qquad (3.5)$$
$$\|y(t) - \bar{y}_q(t)\| < \eta_q \rightarrow fault_q\ identified$$

Example 4: Consider the following system:

$$\dot{x} = -2x$$

$$y = x$$

Let the initial condition be $y(0) = 3$. If the value of the output at $t = 0.5$ is $y(0.5) = 1.5$. Determine the occurrence of fault for the following threshold values:

> **i.** $\eta = 0.2$
> **ii.** $\eta = 0.5$
> **iii.** $\eta = \frac{y}{2}$

Solution

The simulation of the system can be carried out using the following equation:

$$\bar{x}(t) = x(0)e^{-2t} = 3e^{-2t}$$

According to the previous equation, $\bar{y}(0.5) = 1.103$. Now, we apply fault detection for each of the threshold values:

> **i.** $\eta = 0.2 \rightarrow |y(0.5) - \bar{y}(0.5)| = |1.5 - 1.103| = 0.396 > 0.2 \rightarrow$ *fault detected*
> **ii.** $\eta = 0.5 \rightarrow |y(0.5) - \bar{y}(0.5)| = |1.5 - 1.103| = 0.396 \ngtr 0.5 \rightarrow$ *no fault*
> **iii.** $\eta = \frac{y}{2} = 0.75 \rightarrow |y(0.5) - \bar{y}(0.5)| = |1.5 - 1.103| = 0.396 \ngtr 0.75 \rightarrow$ *no fault*

Example 5: Consider the system:

$$\dot{x} = -2x + u$$

$$y = x$$

Where the feedback control system is defined as $u = -3x$. Now consider actuator faults where in fault 1, the control effort is reduced to half, i.e., $u_1 = -1.5x$, and in the second fault, the control effort is reduced to zero, i.e., $u_2 = 0$.

Consider the thresholds of fault identification to be 0.1 for both faults. Furthermore, assume that the initial condition is $x(0) = 8$. Identify the fault under the following conditions:

> **i.** $y(0.5) = 1.45$
> **ii.** $y(1.5) = 0.45$
> **iii.** $y(2) = 0.0005$

Solution

Note that, with the specified control and fault situations, the simulation models are as follows:

$$\dot{\bar{x}} = -5\bar{x} \text{ (normal condition)}$$

$$\dot{\bar{x}}_1 = -3.5\bar{x}_1 \text{ (Fault 1 condition)}$$

$$\dot{\bar{x}}_2 = -2\bar{x}_2 \text{ (Fault 2 condition)}$$

Now we compare in each case:

i. Simulated values are $\bar{y}(0.5) = 0.6567$, $\bar{y}_1(0.5) = 1.3902$, $\bar{y}_2(0.5) = 2.943$. The only condition that matches with $y(0.5) = 1.45$ within the threshold of 0.1 is fault 1 condition. Therefore, in this case, fault 1 shall be identified.

ii. Simulated values are $\bar{y}(0.5) = 0.0044$, $\bar{y}_1(0.5) = 0.042$, $\bar{y}_2(0.5) = 0.3983$. The only condition that matches with $y(1.5) = 0.45$ within the threshold of 0.1 is fault 2 condition. Therefore, in this case, fault 2 shall be identified.

iii. Simulated values are $\bar{y}(0.5) = 0.0004$, $\bar{y}_1(0.5) = 0.0073$, $\bar{y}_2(0.5) = 0.1465$. Note that there are two conditions in this case that match with $y(2) = 0.0005$ within the threshold of 0.1, i.e., fault 1 condition and the normal condition. In this situation, additional mechanism is required to resolve this conflict and declare either normal condition or fault 1 condition.

3.5 RECONFIGURABLE CONTROL AND CONTROL RECONFIGURATION LOGIC

After a fault has been detected and identified, a response is required in order to minimize the effect of the fault. Such a response could be any of the following: turn the system off, use alternate software program (or alternate control law) to govern the system, or use alternate hardware (e.g., a redundant sensor or a redundant actuator) to replace the faulty one, etc. This implies that to enable fault tolerance, a control system must be designed in such a way that at least one of the previous responses is possible. But the availability of a redundant hardware or an alternate control law is not enough. We also need a sound logic that is able to select appropriately among the available choices.

Selection of a post-fault control scheme or hardware is not straight forward. The major challenge is possibility of false alarm or a missed detection. Design of reconfigurable control depends upon the post-fault observability and controllability properties of the system. The controllability is associated with the actuator faults and observability is associated with the sensor faults. Consider the system presented in equation (3.1). In this system, an actuator fault would change the B matrix. Hence changing the eigen values of $(A - BK)$ which govern the stability of the closed-loop system. Also, a change in B matrix may change the rank of the Q matrix, i.e., the controllability matrix. A sensor fault affects the C matrix in the system and hence affects the observability because the rank of the observability matrix O depends upon C. Finally, a fault in the plant affects the A matrix of the system. This is most critical as it affects everything, i.e., closed-loop stability, controllability, and observability. The reconfigurable control design problem can be presented as

Given the post-fault mathematical model of the system, design an observer-based dynamic feedback control law that is able to stabilize the system (or track a reference trajectory), i.e., given:

$$\dot{x} = A_f x + B_f u , \quad x \in \mathbb{R}^n$$
$$y_f = C_f x , \quad y \in \mathbb{R}^p \tag{3.6}$$

Where A_f, B_f, C_f are post-fault matrices depicting the system dynamics and the availability of sensors and actuators. We need to determine K_f and L_f such that the following system is asymptotically stable:

$$\dot{x} = A_f x - B_f K_f \hat{x}$$
$$\dot{\hat{x}} = (A_f - B_f K_f)\hat{x} + L_f C_f e, \quad e = x - \hat{x} \tag{3.7}$$
$$\dot{e} = \dot{x} - \dot{\hat{x}} = (A - L_f C_f)e$$

Note that the previous-mentioned problem is solvable if and only if the following conditions are satisfied:

$$rank\,(Q_f) = rank\,(O_f) = n \tag{3.8}$$

Where,

$$Q_f = [B_f \quad A_f B_f \quad A_f^2 B_f \quad \dots \quad A_f^{n-1} B_f]$$

And

$$O = [C_f \quad C_f A_f \quad C_f A_f^2 \quad \dots \quad C_f A_f^{n-1}]^T = \begin{bmatrix} C_f \\ C_f A_f \\ C_f A_f^2 \\ \vdots \\ C_f A_f^{n-1} \end{bmatrix}$$

In the fault cases where condition (8) is not satisfied, either a redundant actuator or sensor is needed or a pre-determined safe mode is required in order to avoid further damage to the system.

Now, designing alternate control law(s) (for as many pre-determined faults as possible) is only half of the task. The remaining task is to develop an appropriate logic for the selection of post-fault control scheme. As mentioned earlier, a major challenge in designing such logic is that the process of fault detection and diagnosis cannot be assumed as perfect. Consequently, the problem of control reconfiguration logic is a stochastic optimization problem that can modeled as a Markov Decision Process (MDP) [16]. A Markov decision process is specified by states, decisions, cost function, and transition probabilities. For the reconfiguration logic problem, the states are specified as:

$$
\begin{aligned}
S &= \{s_1, s_2, \ldots, s_N\} \\
s_i &= \{P_i, F_i, g_i, h_i\} \\
P_i &= \{p1_i, p2_i, \ldots, pq_i\} \\
pj_i &\in [0, 1] \\
F_i &= \{f1_i, f2_i, \ldots, fq_i\} \\
fj_i &\in \{0, 1\} \\
h_i &\in \{1, 2, \ldots, n1\} \\
g_i &\in \{1, 2, \ldots, n2\}
\end{aligned}
\tag{3.9}
$$

Here, S is the set of states (there is a total of N states). Each state specifies four types of information, i.e., the status of the fault vector F, the status of the fault occurrence probabilities P, the status of the active hardware configuration h, and the status of the active dynamical control law g. There are total q possible faults, the occurrence of a fault (with index j) is indicated by a binary flag fj where $fj = 1$ means that the fault has occurred and $fj = 0$ means that the fault has not occurred. Fault probabilities represent the confidence level in the occurrence flag of a fault. Each fault flag has an associated probability of being true. The probability value ranges between 0 and 1. Note that these probabilities are derived from the probability of correct fault detection [16] which depends upon the reliability of the fault detection and diagnosis scheme and also upon the reliability of the component whose failure is associated with the occurrence of the fault. As discussed earlier, fault tolerance requires redundant hardware, therefore the variable h is included in the state space in order to specify the hardware combination (sensors and actuators) that is currently active in the feedback control system. Note that h may assume any of the $n1$ possible values where each value of h corresponds to a unique combination of hardware components that are active in the underlying feedback control system. Finally, g represents the currently active feedback control scheme, i.e., the state estimation-based dynamic feedback control. According to (9), there are $n2$ possible control schemes.

The set of decisions in reconfiguration problem is given by:

$$
D = \{h1, h2, \ldots, hn1, g1, g2, \ldots, gn2, NOOP\}
\tag{3.10}
$$

The previous equation presents three types of decisions. The first type of decision is to activate a specific hardware configuration among the set of available $n1$ configurations. The second type of decision is to activate a specific dynamic feedback control scheme among the set of available $n2$ control schemes. Finally, the *NOOP* decision is no-operation, i.e., let the things be as they are (this decision is default action of the reconfiguration logic because a change is required only when a new fault is detected).

The cost function is associated with post-fault performance of the feedback control system. In general, the cost function is of the form:

$$J(s) = Z(P, F, g, h) \tag{3.11}$$

The exact definition of the function Z shall depend upon the nature of the application. One way to define Z is to make the cost of post-fault high for wrong selection of control. For example, suppose that control law 3 is suitable for fault 7, the cost function shall be defined in such a way that the cost of selecting any other control law than the control law 3 under the occurrence of fault 7 shall be very high. This cost can also be made proportional to the level of confidence (P) associated with the fault detection. Furthermore, in case of loss of a component for which a redundant is available, the cost of not using the redundant component shall be defined as high. The exact values associated with the cost function shall depend upon the application and the real-world value of the purpose for which the application is working.

The transition probabilities in the reconfiguration problem involve the probabilities of occurrence of faults. Note that the occurrence of some faults may increase or decrease the probabilities of occurrences of other faults. In practice, these probabilities are developed using the statistical data pertaining to the fault cases from the history and also from the reliability of various hardware components involved in the feedback control. Fault probabilities may also depend upon the active hardware configuration and the control law because some control laws may be more aggressive (risky) than the others. Similarly, some hardware configurations may be more (or less) reliable than the others. Mathematically, the state transition probabilities shall be represented as:

$$T(s_j|s_i, d) = T(P_j, F_j, g_j h_j | P_i, F_i, g_i, h_i), \ s_i, s_j \in S, d \in D \tag{3.12}$$

The previous representation of the transition probabilities depends upon the decision because, other than the fault vector and the fault detection confidence, the control law transition and the transition in the hardware configuration depend upon the decision undertaken at the state s_i.

Once the components of the MDP are specified for a reconfiguration problem, the optimal policy is calculated using a stochastic dynamic programming algorithm such as value iteration.

Algorithm 1: Value Iteration

Function: Value Iteration, Input: S, D, J, T, γ, Output: π^*: $S \rightarrow D$

Step 1: Initialize the value of each state $V^0(s) = 0$, $\forall s \in S$, $t = 0$

Step 2: Update the value of all the states.

$$V^{t+1}(s) = -J(s) + \max_{d \in D}\left(\gamma \sum_{s_j \in S} T(s_j|s, d) V^t(s_j) \right)$$

Step 3: Check the convergence.

$$||V^{t+1} - V^t|| < \eta$$

If the convergence is achieved, define $V^*(s) = V^{t+1}(s)$ $\forall s \in S$ and continue to Step 4. Otherwise, update $t = t + 1$ and go to Step 2.

Step 4: Calculate the optimal policy using the converged value function V^*

$$\pi^*(s) = \underset{d \in D}{\operatorname{argmax}}\left(\gamma \sum_{s_j \in S} T(s_j|s, d) V^*(s_j) \right)$$

Note that $\gamma \in (0,1)$ is the user-defined discount factor for the calculation of the policy. Also, the threshold η is defined as a small number usually $\approx 10^{-6}$

Value iteration algorithm is presented in the following.

The output of the value iteration algorithm is an optimal reconfiguration policy. Next, we discuss the overview of the integration of fault tolerance in feedback control in the light of the previous discussion.

3.6 INTEGRATION OF FAULT TOLERANCE WITH FEEDBACK CONTROL

The steps involved in the design and integration of fault tolerance in feedback control have been presented in Figure 3.2. Note that all of the discussion in this chapter has been regarding the faults that can be anticipated in advance. However, some faults may occur that are not anticipated in advance. A brief discussion regarding such faults is presented in the next section.

According to Figure 3.2, the first step is to make a list of faults that may occur during the feedback control operation. Such faults may include full or partial failure of an actuator, drift or bias error in a sensor, or damage to the physical structure of the plant. Once all the faults have been listed, the corresponding

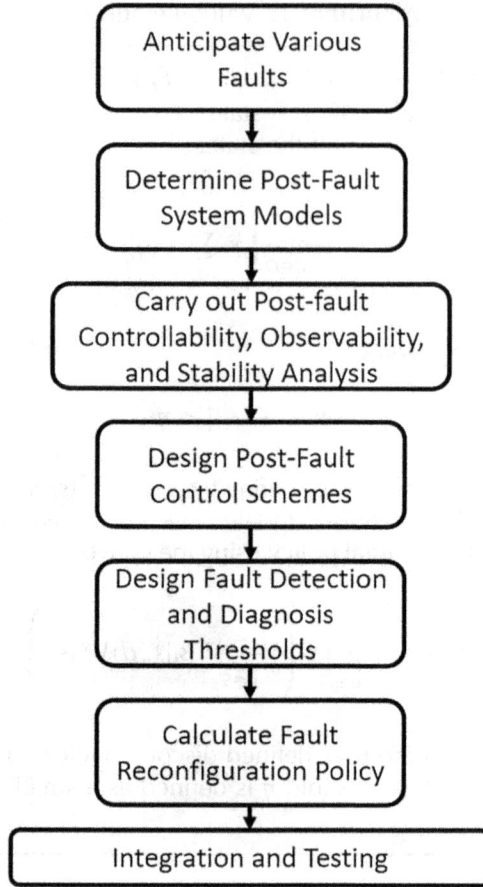

FIGURE 3.2   Steps involved in the integration of fault tolerance in feedback control.

mathematical models need to be developed. Recall that a fault in an actuator causes changes in the B matrix of the system, sensor faults cause changes in the C matrix, and faults in the plant cause changes in the A matrix. After developing the post-fault system models, the next step is to analyze the stability, controllability, and observability properties of the post-fault models. This step is followed by the design of control scheme for each of the anticipated faults. The next step is to design fault detection and diagnosis thresholds. These thresholds may be fixed or adaptive depending upon the nature and dynamics of the faults. Finally, reconfiguration policy is calculated which requires the probabilities associated with the faults and fault detection and diagnosis scheme as discussed in the previous section. Once all the steps are carried out, the fault-tolerant control system shall be integrated and tested before it is deployed. In some cases, arranging for the redundant hardware components and integration of the same within the system is also part of the process.

```
┌─────────────────────────┐
│      Define "safe"       │
│     operating mode       │
└─────────────────────────┘
             ↓
┌─────────────────────────┐
│     In case of unknown   │
│   fault, bring the system│
│       into safe mode     │
└─────────────────────────┘
             ↓
┌─────────────────────────────┐
│   Diagnose post-fault system │
│      behavior using system   │
│ identification or machine learning │
└─────────────────────────────┘
             ↓
┌─────────────────────────┐
│  Redesign feedback control│
│   scheme to cater for post-│
│    fault system dynamics  │
└─────────────────────────┘
             ↓
┌─────────────────────────┐
│     Return to normal     │
│     operating mode       │
└─────────────────────────┘
```

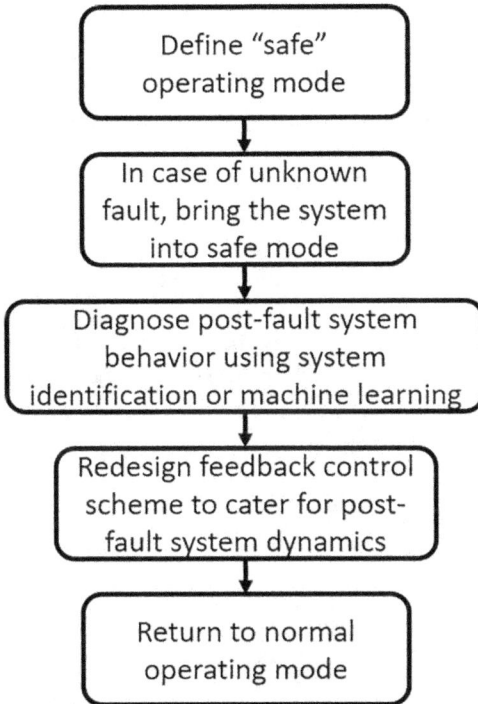

FIGURE 3.3 Process for dealing with unknown faults.

3.7 DEALING WITH UNKNOWN FAULTS

Dealing with unknown faults is quite challenging because not only does it require thorough diagnosis, but it also requires online redesign of the controller. In case of unknown faults, the system may have to discontinue its normal operation and shift into diagnosis mode. In diagnosis mode, some pre-determined safe control commands are executed in order to gather data and response of the post-fault system. From the data (response) gathered, new model of the system is identified and all the necessary analysis is performed. After the analysis, the control scheme is designed for the post-fault system and the system can resume operation. Note that such capability can be incorporated along with the previously discussed pre-calculated fault tolerance. In some situations, machine learning-based approaches can be used to identify the post-fault model of the system. The details of the design are beyond the scope of this chapter; however, a flowchart is provided in Figure 3.3 for understanding of the concept.

3.8 CONCLUSIONS AND SUMMARY

This chapter has presented a methodology for enabling fault tolerance in feedback control systems. The chapter discusses the basic concepts and techniques by assuming an underlying linear time-invariant continuous-time dynamical system.

The overall concept can be adapted (with appropriate modifications) for the nonlinear or discrete-time systems.

Fault diagnosis and detection is nontrivial because sometimes it is hard to differentiate between a fault and a disturbance, and also two or more faults may result in similar system behavior hence making it difficult to pinpoint the issue. Fault reconfiguration is challenging because there are uncertainties associated with the problem such as probabilities of false alarm and missed detection. Furthermore, the computational complexity of the whole process grows rapidly as the number of faults increases or if multiple simultaneous faults are considered.

Dealing with unknown faults has been briefly discussed in the chapter, and the challenges involved in the process have been highlighted. Unlike in the case of pre-determined fault situations, the occurrence of an unknown fault may cause the system to shift to a safe operating mode (disrupting the normal operation of the system). In general, a fault-tolerant control system can be designed to have the capability to deal with both known and unknown faults.

NOTE

1 Two or more columns of a matrix are linearly independent if and only if their linear combination is nonzero for all nonzero coefficients, i.e., c_1 and c_2 are linearly independent if and only if $\alpha c_1 + \beta c_2 \neq 0, \forall \alpha, \beta \in \mathbb{R} \{\alpha, \beta\} \neq \{0,0\}$.

REFERENCES

[1] Zhang, Y., and Jiang, J. "Bibliographical review on reconfigurable fault-tolerant control systems," *Annual Reviews in Control*, 32(2), 229–252, 2008.
[2] Blanke, M., Izadi-Zamanabadi, R., Bogh, R., and Lunau, Z. P. "Fault tolerant control systems—A holistic view," *Control Engineering Practice*, 5(5), 693–702, 1997.
[3] Zhou, D. H., and Frank, P. M. "Fault diagnostics and fault tolerant control," *IEEE Transactions on Aerospace and Electronic Systems*, 34(2), 420–427, April 1998.
[4] Patton, R. J. "Fault-Tolerant Control: The 1997 Situation," *In Proceedings of the 3rd IFAC Symposium on Fault Detection, Supervision and Safety for Technical Processes* (pp. 1033–1055), August 1997.
[5] Zhang, Y. M., and Jiang, J. "Fault tolerant control system design with explicit consideration of performance degradation," *IEEE Transactions on Aerospace and Electronic Systems*, 37(3), 838–848, 2003.
[6] Chamseddine, A., Noura, H., and Ouladsine, M. "Sensor fault detection, identification and fault tolerant control: Application to active suspension," 1-4244-0210-7/06 2006 IEEE.
[7] Nasir, A., and Atkins, E. M. "Fault Tolerance for Spacecraft Attitude Management," In *AIAA Guidance, Navigation, and Control Conference*, Toronto, Ontario, Aug. 2–5, 2010 (AIAA-2010-8301).
[8] Rago, C. et al. "Failure Detection and Identification and Fault Tolerant Control Using the IMM-KF with Applications to the Eagle-Eye UAV," In *Proceedings of the 37th IEEE Conference on Decision and Control*, Tampa, Florida, USA, December 1998.

[9] Zhang, Y. M., and Jiang, J. "Integrated active fault tolerant control using IMM approach," *IEEE Transactions on Aerospace and Electronic Systems*, 37(4), 1221–1235, October 2001.

[10] Blanke, M., Michel Kinnaert, J. L., Staroswiecki, M., and Schröder, J. *Diagnosis and fault-tolerant control*. Vol. 2. Berlin: Springer, 2006.

[11] Ding, S. X. *Advanced methods for fault diagnosis and fault-tolerant control*. Berlin: Springer, 2021.

[12] Amin, A. A., and Hasan, K. M. "A review of fault tolerant control systems: advancements and application," *Measurement*, 143, 58–68, 2019.

[13] Van, M., and Ceglarek, D. "Robust fault tolerant control of robot manipulators with global fixed-time convergence," *Journal of the Franklin Institute*, 358(1), 699–722, 2021.

[14] Zhu, G., Ma, Y., Li, Z., Malekian, R., and Sotelo, M. "Event-triggered adaptive neural fault-tolerant control of underactuated MSVs with input saturation," *IEEE Transactions on Intelligent Transportation Systems*, 23(7), 7045–7057, 2021.

[15] Hu, Q., Li, B., Xiao, B., and Zhang, Y. "Spacecraft attitude fault-tolerant control based on iterative learning observer and control allocation," In *Control allocation for spacecraft under actuator faults*. Singapore: Springer, 2021, pp. 133–155

[16] Nasir, A., Atkins, E., and Kolmanovsky, I. "A mission-based fault reconfiguration framework for spacecraft applications," In *Infotech@ Aerospace 2012*, p. 2403, 2012.

Part 3

Functional Reverse Engineering of Controller

4 Design and Development of 3-Axis Wire Arc Additive Manufacturing Machine

Functional Reverse Engineering Approach

G. Hussain, Osama Al-Jamal, Adeel Ghafoor, Abrar Farooq, Usman Zahid, Iftikhar Ahmad, and Muhammad Iqbal

4.1 INTRODUCTION

The concept to design and fabricate was inspired by an attempt to obtain a WAAM machine with 3-axis capability for advanced production methods in the industry. 3-axis machines can manufacture more complicated geometries faster and for less money. These machines increase productivity by lowering set up times, cycle times, and gun wear while also boosting capabilities. Time and money are saved as a result of the shorter set up times. WAAM technology is used in various industries, including aerospace, automotive, medical, architectural, and energy [1]. Programs that demand low production, high design complexity, and the capacity to alter design frequently are examples of these. WAAM is seen as a promising alternative for producing metallic materials and is well-known for its high-quality parts. The precision of 3-axis Computer Numeric Code (CNC) WAAM machine is also improved because it requires minimal setups; additional configurations will just result in more errors [2].

For the 3-axis CNC machine development, a three-axis CNC gantry machine was modeled in SolidWorks and the model was fabricated. Arduino and Raspberry Pi were used to automate the machine with a single piece of software. The motive is to produce a collection of rules that the industries may use to suit their production needs. With technology advancing at a rapid pace, it is vital for business firms to upgrade their production processes on a regular basis [3].

DOI: 10.1201/9781003376620-7

55

The chapter is comprised of four sections. Section 4.1 describes the 3-axis CNC WAAM machine and its operation. It also identifies the key issue statement that the project aspires to solve throughout the course of its life. This section also discusses everyone's contributions to the project's overall success. Section 4.2 presented a detailed literature review. The design element of the project is covered in Section 4.3. It details the entire procedure for doing a design analysis of a model created with CAD software. This section also shows the CAD model as well as the analysis, calculations, and simulations performed. Section 4.4 reports the development and experimentation.

4.2   3-AXIS WAAM MACHINE

Additive Manufacturing (AM), also known as the "third industrial revolution" due to its ability to rework the manufacturing market, is the manufacturing process in which a material is added to each layer to make a product. Additive Manufacturing processes extract data from a computer-assisted file (CAD) that is later converted to a stereolithography (STL) file. In this process, a drawing made on CAD software is measured in a triangle and cut out, containing details of each layer to be printed [4]. AM is extensively used in aerospace, automotive, medical, architectural, and energy sectors. These include programs that require low production, high design complexity, and the ability to change designs frequently. However, AM still needs significant development in terms of design, materials, new technologies, and machines [5].

The aerospace industry has shown interest in these technologies because of their ability to fabricate metal parts directly using materials such as titanium (suitable for aircrafts) and the ability to easily manufacture complex, high-performance products without any tools [6].

Ding et al. [7] investigated the components of the aircraft with thin-walled structures of different thicknesses and included several sections using the method of AM based on arc welding.

Finally, as reported by [8], AM technologies have increased the productivity of the aerospace industry by reducing production time by 30–75%, reducing one-time product costs by up to 45%, and reducing the cost of producing small parts by 30–35%.

4.3   WAAM MACHINE COMPONENTS

4.3.1   MECHANICAL SYSTEM

The mechanical system is set up in such a way that linear bearings and guide rods are used to move the three linear axes. Each axis has a stepper motor attached to it, which is the source of motion acted upon by the control signal created by the circuit. Each linear motion stepper motor was connected to the screw rod, which, with the help of a coupling bush, carries the nut. The screw rod and nut assembly are responsible for turning the stepper motor's rotating motion into linear motion. Each axis's linear motion is smoothly carried away by the linear bearing and guide rod assembly connected to it, which can carry load and linear motion at the same time. The rotation of the stepper motor is directly controlled to create controlled motion in each axis. Direct control of the stepper motor's speed by providing the needed control signals can also be used to regulate the speed of motion in each axis.

4.3.2 ELECTRICAL SYSTEM

The main components of electronic system are comprised of:

- Power supply
- Microcontroller
- Stepper motors
- Stepper motor driver shield.

4.3.3 POWER SUPPLY

The power supply, which converts AC electricity to DC voltage and distributes appropriate voltages to related devices such as stepper motors, driver shields, and microcontrollers, is at the core of the CNC system. Most microcontroller boards and driver shields work with 5 volts, but stepper motors need 12 volts or more, depending on how big they are and how they are set up.

4.3.4 STEPPER MOTOR

A stepper motor is one of the main components of the electronic system of a WAAM machine. It converts rotational motion to linear motion for the 3 linear axes, that is, X, Y, and Z. The stepper motors are used according to the magnitude of the design torques calculated for each axis.

4.3.5 MICROCONTROLLER

A microcontroller is a computer having one or more CPUs (processing cores), memory, and programmable input and output peripherals. A modest amount of RAM, as well as program memory in the form of ferroelectric RAM, NOR flash, or OTP ROM, is frequently incorporated on the chip. Microcontrollers, in contrast to the microprocessors found in personal computers and other general-purpose applications, are developed for embedded applications and consist of a number of discrete chips. The Raspberry Pi is an example of a microcontroller with advanced processing capabilities that can execute complex tasks simultaneously. By smoothing motor action and decreasing jerks and vibrations, the Raspberry Pi's pulse-width modulation programming capability benefits WAAM machine design experts even more. PWM (pulse-width modulation) or PDM (pulse-duration modulation) is a technique for converting a pulsing signal into a message [9].

4.3.6 SOFTWARE

A modern WAAM system's software must have at least three major programs: a part program, a service program, and a control program. The part program can be created using CAM (Computer-Aided Manufacturing) software. The part program is checked, edited, and corrected using the service program. It usually comes with a user interface that makes it simple to use the equipment. The control program takes the part program as input data and generates signals to drive the stepper motors' axes of motion. It has interpolation, feed rate control, acceleration and deceleration operations, as well as position counters that show the current axis position. A built-in algorithm has been developed to read control programs for any sort of partial program. This pre-programmed algorithm is written in Python and stored on a microcontroller. In today's WAAMs, the Raspberry Pi is employed [10].

4.3.7 INFERENCES

In the wake of going through the most recent innovations utilized and plans made of WAAM, the final design was crafted using SolidWorks, which revolved around the ideas abstracted from the previous working models. The x-axis is now attached to the z-axis, so our final design has all three axes. Motors are used, where two are for the linear motion along x and one each for y and z. Ball screws and linear rails are used for translation. Motor is selected based on the weight of the axis assembly. These weights convey to us the required rating of the motors. The software portion and development of the graphic user interface will be done using the language Python, which has a very clean and easy-to-read syntax. The hardware used to support this language must be a single-board computer such as a Raspberry Pi or Arduino.

4.4 DESIGN AND ANALYSIS

3-Axis machining constitutes three translational axes in the x, y, and z directions. The manufacturing process of complex parts has been revolutionized by the capabilities of multi-axis machining. Conventional machining processes are slow, inflexible, and limited in their functions and capabilities as compared to multi-axis machining such as 3-axis milling machines. Hence, by overcoming the constraints of traditional machining, it can provide the advantage of manufacturing complex parts with increased accuracy and higher surface finish with reduced setup time. Consequently, enhancing machine accuracy has been one of the main focuses of research on 3-axis machine guns in the past [11].

In the design and manufacturing industry, the NC concept refers to the automation of machine guns (MIGs) that are operated by abstractly pro-grammed commands encoded on storage media such as G-codes. The first NC machines were built based on existing guns that were modified with motors that moved the controls to follow points fed into the system on paper tape. The early servomechanisms were rapidly improved with analog as well as digital computers, creating the modern computer numerically controlled (CNC) machine tools. A CNC machine includes several types: 2-axis CNC machine, 3-axis CNC machine, 4-axis CNC machine, 5-axis machine, etc. The number of the axis of a CNC machine implies the capability of the controller of the machine to interpolate simultaneously. If the axis number increases, the machining efficiency, effectiveness, and accuracy will increase. However, it requires more complex techniques in tool path generation and process programming with advanced controllers such as the Raspberry Pi. The 3-axis WAAM CNC machine has been proven to be the most efficient tool for fabricating products of complex geometry, which may include several free-form surfaces. Integration of 3-axis CNC systems with CAD-CAM systems has revolutionized industrial automation. The following list shows the main components of modern CNC.

4.5 DESIGN METHODOLOGY

This work aims to provide the best combination of cost, quality, and esthetics to develop a prototype CNC WAAM machine with 3-axes that gives more accessibility, intricacy, and even more accuracy over the fabrication of complex parts, which is the reason why WAAM CNC machines are highly in demand in the aerospace industry [12]. The era of control is in function and the availability of new computer technologies such as low-cost open-source hardware, such as Arduino and Raspberry Pi microcontrollers, makes machines programmable. The web's availability of a wide range of ready-to-use software languages, such as Python and C++, has significantly reduced prototyping and development times. Furthermore, by utilizing such readily available machinery and putting it into desired motion, the demands for low-cost CNC WAAM Machines from small-scale to large-scale industries will be met with reduced costs and increased functionality [13].

From the viewpoint of machine control, such tools may be quite adequate for the development of low-cost models of CNC machines. The development of a prototype 3-axis CNC WAAM machine is presented to be easily operable, easy interface, flexible, and portable.

This project has been classified into the following modules for successful execution:

- Mechanical Design and Analysis Module
- The Electronics and Control Module
- Software Module.

The system works in a manner such that the machine operator inputs the G- and M-codes into the GUI (Graphic User Interface), which then sends the G- and M-codes to the software, which interprets the code and sends the signals to the electronic driver. The driver then runs the motors, which are connected to the lead screws that are responsible for the movement of the gun (MIG) [14].

TABLE 4.1
Material Selection for Bed

Material	Density g/cc	Thermal Conductivity (W/m-K)	Price/kg (PKR)	Melting C⁰
Cast Iron	5.54–7.81	11.3–53.3	150–300	1200–1310
Aluminum	2.7	167	300–400	582–652
Stainless Steel	8	16.2	600–700	1400–1420

4.6 MATERIAL SELECTION FOR BED (ANALYTICAL HIERARCHY PROCESS)

The density of the machine was our priority, and we needed to make sure that the machine would be easy to carry and light in weight. Material costs were determined and needed to be as modest as possible. Stiffness was not our priority because the machine may get damaged under heavy vibrations. The material should be durable with considerable strength. Damping materials were thought to reduce or eliminate the damaging forces caused by mechanical or electrical energy. Thermal stability is the measure of workability under high temperatures (Table 4.1).

The data collected for the selection was from literature review of already developed 3-axis machines, and specific properties were obtained from ASM standards.

4.7 GEOMETRIC MODELING AND DESIGN

To give our thoughts and ideas a physical meaning, the modeling and design phases are essential. It is so that we can brain-storm on the research study to amplify our thought process for the collection of required data. The developed ideas through our research studies and explorations led us to our main objective.

4.7.1 CAD Design

The 3-axis CNC machine was modeled after using the previous studies of structures and machines and by using the concept of a 2D plotter. Selection of material and by considering design limitations and specifications. The software used for this purpose was Solidworks 2019. The CAD representations of the final model are shown later. Figure 4.1 shows the final assembly of the 3-axis CNC WAAM machine (Figure 4.2).

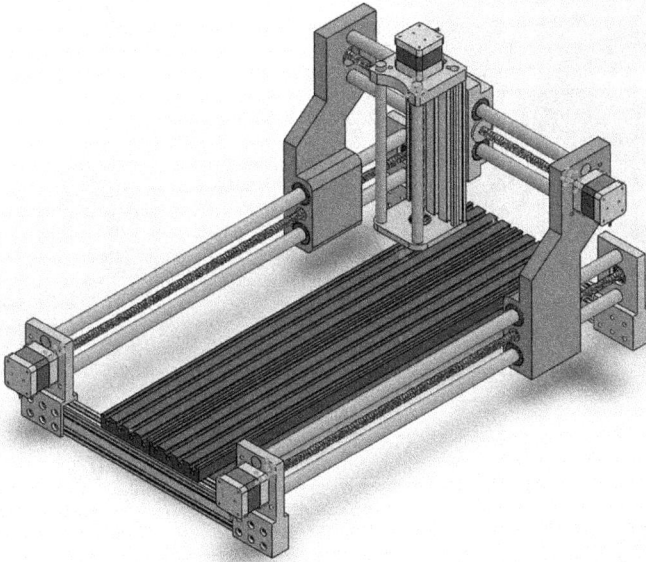

FIGURE 4.1   Conceptual 3-D CAD model.

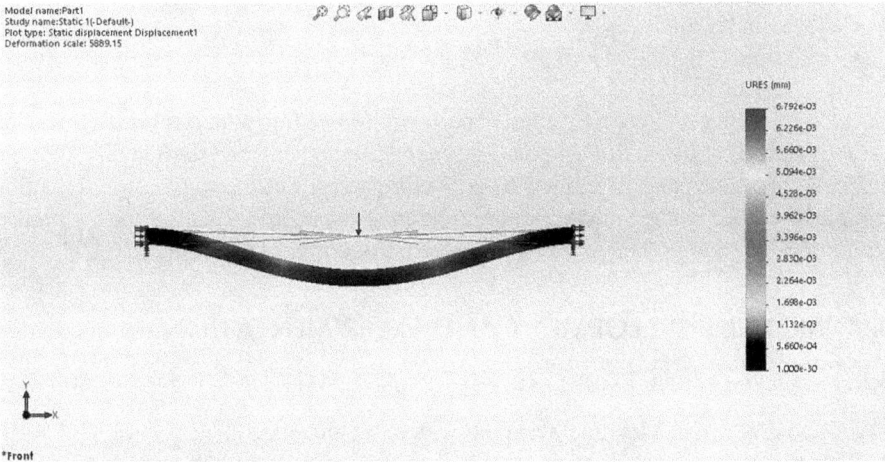

FIGURE 4.2   Bending analysis of shaft.

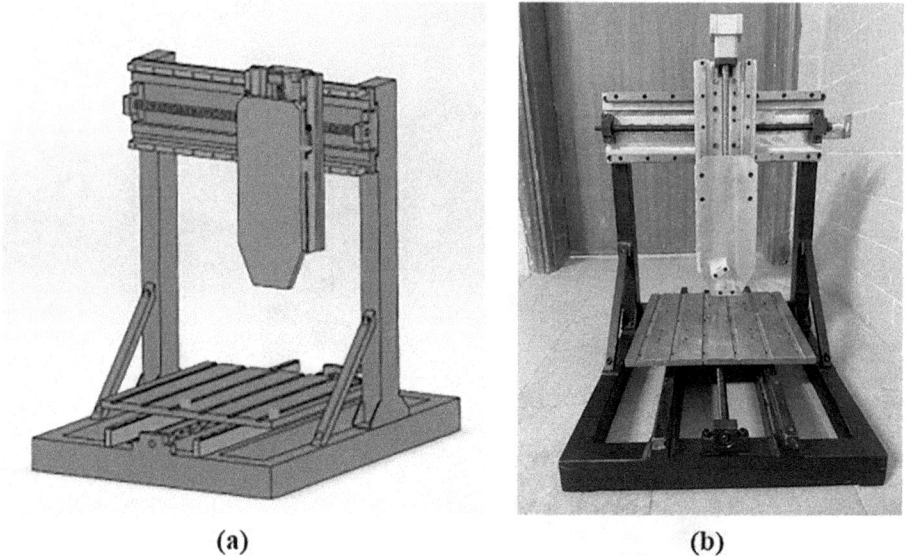

(a)   (b)

FIGURE 4.3 (a) Final detailed CAD model. (b) Picture of the fabricated model.

4.8 ANALYSIS

4.8.1 BENDING ANALYSIS SIMULATION

- Load applied at center of gravity
- Mass of components in z-axis supported by x-axis shafts
- Two shafts in x-axis
- Maximum deflection = 6.792e-03 mm.

As shown in the previous figure, after performing bending analysis on a shaft of z-axis, which is in a critical position, check whether the shaft will undergo deformation when load is applied by a z-axis gantry. So, the maximum deflection computed earlier using Solidworks simulation software is 6.792e-03 mm, which is within the safe range, hence the shaft selected is correct (Figure 4.3).

4.9 MODEL DEVELOPMENT AND EXPERIMENTATION

4.9.1 DEVELOPMENT PROCESS OF THE WAAM MACHINE

The 3-axis WAAM Machine was fabricated by material selection of the body, dimensionally accurate printing of blocks with respect to the industrial scaled measurements, selection of stud (MXX x 2 bolt) for allowance of movement in all axes, selection of adjustable shoes to limit vibrational effects and to level the machine horizontally, construction of x-axis, y-axis, and z-axis assemblies, coupling of stepper motors with the designated axes, and checking of alignment to avoid jamming of any axis during movement.

The following stepper motor configuration was finalized for the model considering a stepper motor NEMA 23 BH110SH 201, which has a holding torque of 1.26 Nm, which is enough for all axes as shown in the static analysis, with a maximum factor of safety of 2.63 at X, Y, and Z axes.

- 01 NEMA 23 in Y-axis
- 01 NEMA 23 in Z-axis
- 01 NEMA 23 in X-axis.

Finally, all the axes were assembled together on the base table by screwing together all the components collectively, and electronic and miscellaneous circuitry was connected with the mechanical body.

4.9.2 FABRICATION OF THE WAAM MACHINE

The machine manufacturing process consists of the following parts:

4.9.2.1 Bed

We have used one bed connected with a linear guide. One is basically for the base plate purpose of WAAM, and all the other assemblies are for supports and lead screw mechanism joining, so that the ball screw is not directly attached to the base plate to interrupt any type of the WAAM production. Mild steel (AISI 1020) plate dimensions (350 × 350 mm) and thickness of 10 mm were used.

4.9.2.2 Linear Guide

The three sets of linear guides are used with a ball screw (pitch of ball screw = 5 mm) for the translation motion to be carried. The reason why linear guidance was used is that it has a higher load-carrying capacity with minimum vibrations and smooth motion.

4.9.3 MOTION MECHANISM OF WAAM MACHINE

Two smooth sliders and a ball screw make up the sliding mechanism. Two ball bearings are mounted to the rectangular brackets and support the ball screw. During the rotation of the ball screw, the ball bearing provides support and reduces friction. Clamps are fastened to brackets and support the slider rods. With the help of plates, slider rods are connected to ball screws. These plates keep the ball screw housing from turning over, as well as the steel plate beneath it, to which the aluminum plate is attached. A supplier for the sliding mechanism has yet to be found.

4.9.4 MOTOR HOUSING

The motor housing consists of a 6 mm thick steel plate which is supported by four pins. These pins are connected to brackets with the help of 5 mm diameter screws. The motor is attached to a steel plate and is also coupled with the ball screw by the help of a flexible coupler.

4.10   INTEGRATION AND INSTRUMENTATION

4.10.1   CIRCUITRY OF THE SYSTEM

The connection between the stepper motors and the axes is quite complex. As the stepper motors were controlled using the Arduino UNO, they had to program the stepper motors through drivers. As the machines were to be produced on an industrial scale, stepper motors with more torque were used. To program and control the stepper motors, the Arduino UNO and drivers (Tb-6560) were used. This driver eases interfacing between the Arduino UNO and stepper motors and is cheaper than other options.

The figure shows a schematic diagram of the electronic setup of the WAAM machine made on Proteus. The electrical interference is because of the limit switches for each axis, and they might give the wrong signal to the raspberry pi, so 4n35 opt couplers are added to reduce electrical interference for accurate motor stepping. The CLK-, EN-, and CW- are grounded. And CW+ and CLK+ are wired into the Raspberry Pi. The circuit is built on Proteus 8.6. The electronic design has been simulated using the Proteus Design Suite software [15] (Figure 4.4).

4.10.2   CONNECTIVITY OF THE COMPONENTS

Coding is done using the Python language and the Raspberry Pi sends a signal to the motor driver (DM 862) to turn on and off the motor with the help of limiting switches. The motor is linked through limiting switches with the help of a Raspberry Pi. The purpose of the motor driver is to provide more voltage to the motor than the Raspberry Pi cannot. Each axis has its own limiting switches and is controlled through the Raspberry Pi [8] (Figure 4.5).

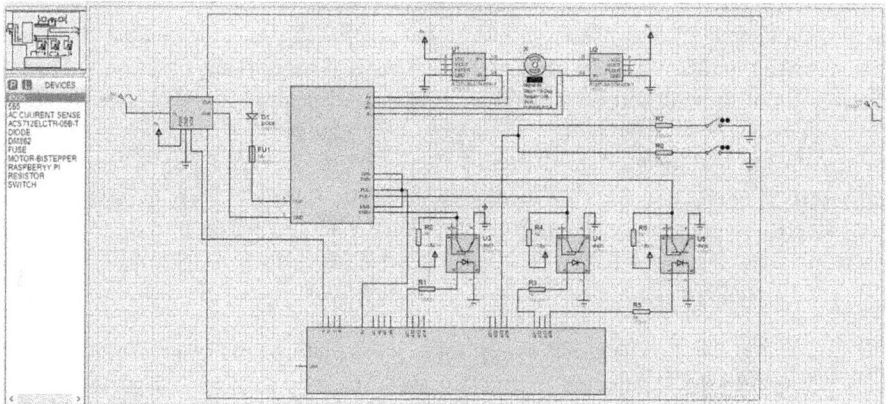

FIGURE 4.4   Circuitry of the system.

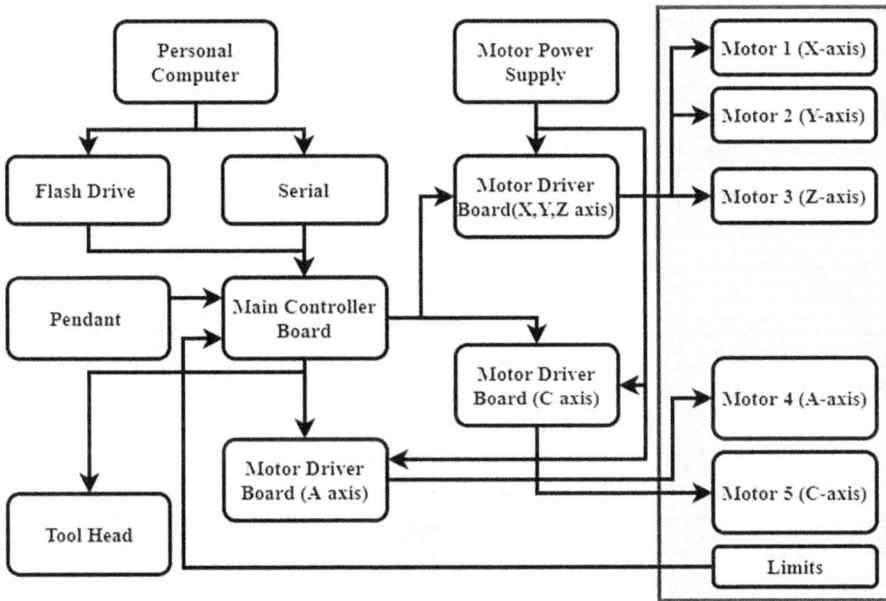

FIGURE 4.5 Connectivity of the components.

4.11 DEVELOPMENT OF THE SOFTWARE

4.11.1 SCOPE DEFINITION AND FORMAT CLASSIFICATION

The process of developing the program, i.e., software, of our computer numeric control machine is initiated from the basics of CAD/CAM.

Having studied the G-code and M-codes initially in our courses, there was clarity as to where to start from. The following three-step procedure was implemented for the interpreter development:

1. Command Identification
2. Command Formation
3. Command Execution.

4.11.2 COMMAND IDENTIFICATION

This step involves setting a format for our numeric control part program. This is done to have a standard NC part program structure that the controller understands. This NC program is made up of several commands that are put in a specific sequence. This series of commands is input into the microcontroller, which must be programmed to interpret and execute these commands.

The microcontroller (an Arduino UNO being used in our project) performs this action by generating the required waves for each of the motor drivers.

The commands that are input into the controller are called "blocks." These blocks hold multiple commands that are grouped together to allow the machine to execute a certain process.

These commands and their functions are identified by us through the standards set by ISO-6983-1:2009.

We must then feed our controller the command specifications and the sequence and format of words. The following is the format classification used for the logic design of an interpreter [16].

4.11.3 COMMAND FORMATION AND EXECUTION

The microcontroller we use in this project is the Arduino UNO, which provides us with a Microsoft platform along with a good share of GPIO pins, i.e., the input–output pins required to connect the motor drivers and other components with. What is crucial for giving life to our final mechanical assembly is the efficient programming of software that interprets the subset of G-code and provides us control over the stepper motors. These stepper motors move quickly; hence, the WAAM machine will run smoothly and without any mistakes.

It is important for us to feed detailed and error-free algorithms to the microcontroller for it to understand how to proceed with the command formation and execution phase.

Dedicated algorithms are needed for each G-Code command function and a dedicated algorithm for checking the syntax of our part programs [17]. Moreover, Checking and simulation windows should be there to allow for the identification of possible mistakes.

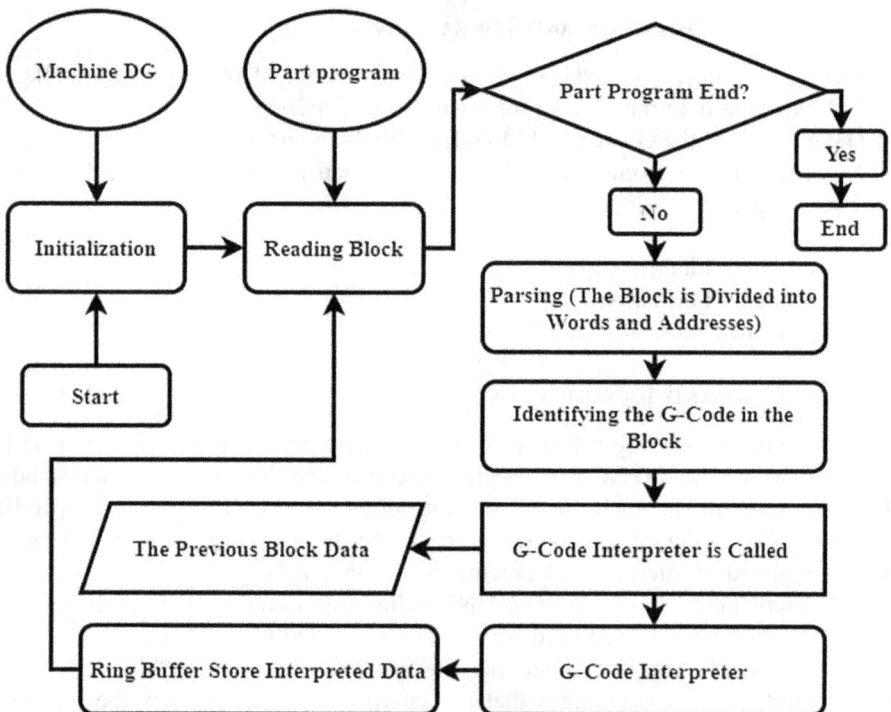

FIGURE 4.6 Basic mechanism of G-code running and interpretation.

Another step of command formation is specifying the possible movement of our axes.

The following are the stages involved in the different combinations of ace movement. The respective algorithms must be fed into the code for the microcontroller to identify the command and send the desired pulses to the motor mechanism.

Stage 1: Individual running of each axis (X, Y, Z).

Stage 2: Combinational movements of two axes (X-Y, X-Z, Y-X, Y-Z, Z-X, Z-Y).

Stage 3: Combinational movements of three axes (X-Y-Z).

4.11.4 ALGORITHMS

Multiple algorithms were generated along the course of this project. The algorithms were given the form of flowcharts, and in the initial phase of the project, these flowcharts were meant to display the basic mechanisms of our NC part program code [18,19] (Figures 4.6 and 4.7).

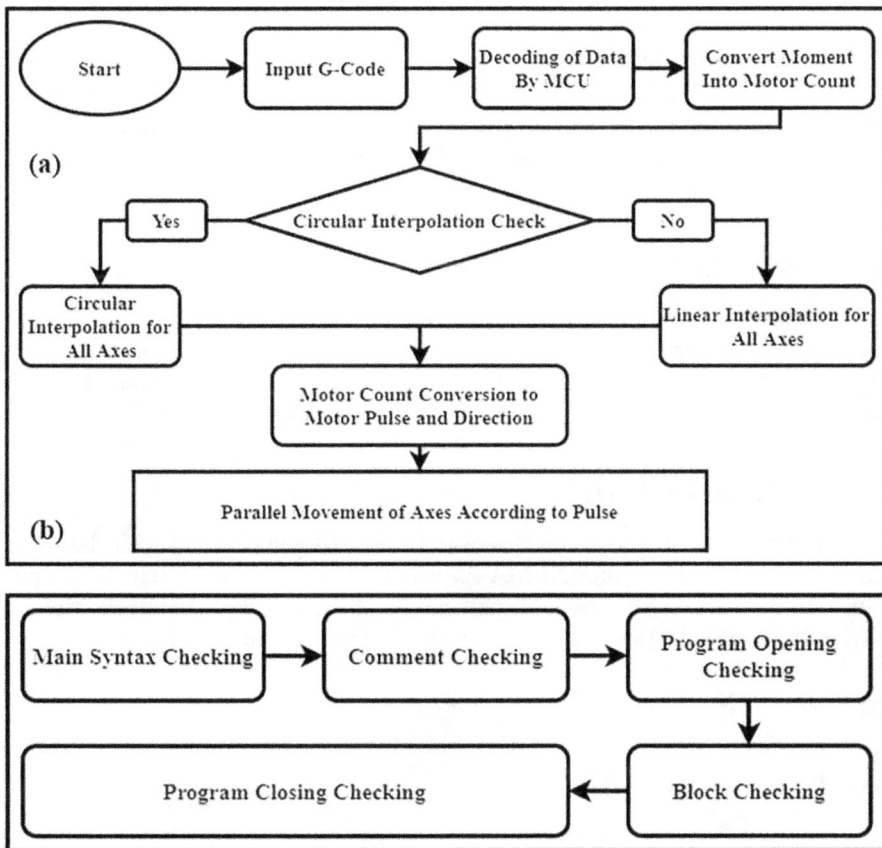

FIGURE 4.7   (a) Decoding of data, converting motor count to motor pulse and direction for the axis movement to take place; (b) syntax checking hierarchy.

4.12 CONTROL SYSTEM

The instruments used in the control system communicate with the user through a microcontroller. The process is simple; the motor receives a signal and moves a specific distance at a specific speed. As the weld torch starts, motion welding also starts at the same time. But the little complexity we need is to precisely synchronize the motion of both the weld torch and the base so that after every pass of the torch the base is decremented by the same amount. So, the user specifies an offset for the base after some seconds equal to the time required for the torch to complete one path. A Graphical User Interface needs to be developed in the C language to make the control more user-friendly. So, through this, the user can enter values for speed and offset length for the base

4.13 TESTING/EXPERIMENTAL PROCEDURES

In the case of 3-axis WAAM machines, the improvement of their motion accuracies is a crucial demand in the market. The testing is to be done on current ISO standards (ISO 10791). The best possible experimental procedure for testing is the printing test.

Various Arduino UNO codes with varying parameters, changing the number of steps, delay, sequence direction, and mode of interpolation each time were created. First, interfaced the code with one motor and then gradually tested different codes on all the three motors. In this way, we found out the threshold of all the motors that were used in the 3 WAAM machines.

4.14 RESULTS AND DISCUSSION

After performing different analyses, we have carried out structural analysis in detail to ensure that our structure withstands mechanical stresses and deformations. For this, shear stresses and deformation were observed in all three dimensions at critical locations.

4.14.1 Factor of Safety

The stepper motor configuration that was finalized for the model with the consideration of the stepper motor NEMA 23 was holding a torque of 1.26 Nm, which is enough for all axes as shown in the static analysis in the previous chapter.

The factor of safety considered for each axis component is as follows:

• 2.63 for X, Y, and Z axes components [maximum factor of safety].

4.14.2 Displacement Analysis

Different modern electrical and mechanical components were used to enhance the accuracy and precision of the fabricated machine. Two different CAD models were designed, and a comparison was made between the two designs. Linear rods were used in the conceptual design and upon computing FEA analysis, it was found that there was a deflection of approximately $6.79 * 10^{-3}$ mm, which may have damaged the performance of the machine. To cater to this issue, linear rails were used in all three axes, which provide smooth motion with no deflection, improving the overall machine performance.

- Load applied at center of gravity
- Mass of components in z-axis supported by x-axis shafts
- Two shafts in x-axis
- Maximum deflection = 6.792e-03 mm.

4.15 ANALYSIS AND DISCUSSION

The 3-Axis Wire Arc Additive Manufacturing Machine is a pillar of many assembling and manufacturing plants.

4.15.1 SAVE TIME

One of the clearest points of interest of a 3-Axis Wire Arc Additive Manufacturing Machine is the decrease in time it takes to finish a venture. It is an additive technique to manufacture a metallic part, having better accuracy than traditional subtractive manufacturing and employing fewer workers to operate the machine. Moreover, little or no maintenance is required for the machine [20]. Furthermore, a single command is enough to operate the machine, and it can continuously work for 24 hours without human intervention. It produces a part with a smoother surface finish.

4.15.2 SAVE COST

Using a 3-axis WAAM machine can also reduce overall production costs per part. Reduced labor saves money and, more importantly, saves approximately 20% of material when using additive technology. As material is saved, this results in a lower cost of material used per part.

4.15.3 COMPATIBLE WITH CAD/CAM

The technology behind computer-aided design and computer-aided engineering has advanced by leaps and bounds over the past decade or two. Previously, there was only limited software support for 3-axis machines. Simply STL. CAD file is used to generate G-codes of the CAD model, and G-codes are used to control the motion of the stepper motor.

4.16 IMPACT AND ECONOMIC ANALYSIS

4.16.1 SOCIAL IMPACT

The proposal we put forward impacts the community from multiple points of view. Prior to the commercialization of the 3-axis WAAM machine, all unpredictable parts that could not be manufactured on a single machine were moved from machine to machine to be finished. For example, in the aerospace industry, rotor blades were primarily assembled through milling and then turned for precise shape; this arrangement lacked accuracy in the final item. Along these lines, with our machine, convoluted configurations are conceivable for machines that, from the outset, required a progression of uncommon apparatuses and arrangements. This thusly builds the accuracy, just as with every arrangement change, error increments.

We are furnished with greater adaptability in manufacturing by the WAAM machine. We can manufacture all the obvious shapes by utilizing this machine. The three axes make the proficient manufacture of certain parts conceivable. The part completion is increased as it is less influenced by vibrations due to the strong platform.

Aside from that, our project paves the way for the mechanical-level creation of any 3-axis machine conceivable. This means many openings for work for the professionals and specialists, including the engineering department. Previously, if one section required several machines to produce, it now only requires one machine with three axes. This reduces the amount of time required for machining as well as the amount of power used; this effect will also be investigated in later stages of this chapter [21].

4.17 SUSTAINABILITY ANALYSIS

Keeping in mind the sustainable development goals, the focus is on improving energy efficiency and reducing waste with a high-quality 3-axis WAAM machine. A sustainability evaluation method for process planning for machine tools considering energy-efficient control strategies has been developed [22]. First, four energy-efficient control strategies for WAAM machines are constructed to reduce their energy consumption. Second, a bi-level energy-efficient decision-making mechanism using random forests is established to select appropriate control strategies for different occasions. Third, three indicators are adopted to evaluate the sustainability of process planning under the consideration of energy-efficient control strategies, i.e., energy consumption, relative delay time, and machining costs. Finally, a pedestal part manufactured by a 3-axis WAAM is used to verify the proposed methods. The results show that if energy-efficient control strategies are used, energy use can be cut by 25%.

3-axis WAAM machines inaccessible in the market are unmistakably more costly than the anticipated financial plan at the culmination of this project. Our undertaking is the initial phase in making a modern standard 3-axis manufacturing machine. However, the machines accessible in the international market vary from 50 to 60k USD; therefore, our determined financial plan for consummation demonstrates that it will be profitable. The machine was controlled by utilizing the Arduino UNO, which will be associated with a screen and input devices. All of these segments are also suitable for use under normal conditions [23].

4.18 ENVIRONMENTAL IMPACT

WAAM machining these days is widely utilized in the public arena. In any case, do you realize what number of contaminations WAAM machining will produce while working? Do you know the classifications of contamination WAAM machining will deliver altogether?

Noise is the sound delivered by the sound body when it is making arbitrary vibrations. Sound is created by the vibration of an object and propagates in a specific medium in the form of waves [24]. The supposed noise pollution has been brought about by man since the modern revolution. The creation and utilization of different mechanical hardware has brought thriving progress to humankind, and yet it has delivered to an ever-increasing extent and more remarkable noise.

The primary drive arrangement of WAAM machines, for the most part, depends on linear guides to finish the transmission. Therefore, the linear guide transmission is one of the fundamental noise sources. Furthermore, the assembly, preload, concentricity, lubrication conditions, and the measure of burden following up on the bearing all affect the noise. The noise of the bearing is also affected by how much the bearing itself differs from what was planned when it was made.

In the procedure of WAAM machining, the utilization of MIG/TIG torch will cause carbon emissions, but the harm is not great. The metal residue particles delivered by this method are bigger, difficult to float, less destructive to inhalation, and do less damage to individuals' bodies than noise pollution.

Gear oil, hydraulic oil, turbine oil, compressor oil, and other oils are commonly used in the industry. After the oil has been in the lubrication system for a while, the oil is influenced by the ambient temperature difference, light, heat, and oxygen. It is influenced by the shearing action to create mechanical impurities. It will cause thermal decomposition, oxidation, and polymerization reactions for a long time to produce oxides, colloids, and asphaltenes dissolved in the oil, which will make the oil turbid, darken, and even darken, causing flashing. The point and kinematic viscosity decrease, moisture and mechanical impurities rise, and oil contamination requires replacement of new oil to ensure the normal operation of the lubrication system. When turning parts, most of them will add alkaline working fluid, which will cause pollution when exposed for a long time [25].

4.19 HAZARD IDENTIFICATION AND SAFETY MEASURES

WAAM machining is a perfect and natural, amicable method of manufacturing which fundamentally runs on power and has no fume exhaust that could influence the earth and environment. The high-impact engines are applied to administer the movement of the axes in machine sync perfectly. The earth is thus completely liberated from commotion deterioration.

We have put together safety measures which need to be taken into precaution:

- Until the administrator is in operation, ensuring the doors are locked with the shutter closed, the WAAM machine will not begin any operation.

- The client ought to allude to the quality directions of WAAM machining before making any effort.
- Electric connections should be made with caution because the machine, once operational, will cause harm not only to the engines, but may also cause issues and influence the administrator. A cut-off power electrical switch is used to unravel this problem.
- The expansion of breaking point switches goes about as a security defender and a crisis switch. When both 3-axes pass their pre-decided position, the switches kill the engine and prevent it from pushing further forward.
- Because the machining activity takes place inside the closed room, the fillings cannot harm the spectators or administrators.
- As it will be warm, it is advised to wear gloves when the manufactured part is brought into contact.
- For protection from heat, wear goggles.

4.20   CONCLUSION AND FUTURE RECOMMENDATIONS

4.20.1   CONCLUSION

In conclusion, the chapter mainly summarizes some key concepts of CNC and design considerations of WAAM machine tools mentioned in the literature. These concepts are implemented in the whole design process of this project. Several components will be chosen based on the reasons discussed. The controller architecture used in the final stage follows the concepts of the WAAM software structure as discussed earlier. The literature provides clear background knowledge and guidance for the development of a prototype of a 3-axis WAAM machine.

The design methodology is discussed in which the mechanical parts used are discussed and how the linear and rotational motions are obtained. After this assembly of mechanical parts, load and torque calculations are performed to obtain a suitable result. Analysis of critical parts in mechanical assembly is discussed, and a final CAD model is attached which is to be manufactured. Also, the concepts of design are discussed.

The circuitry still has room for improvement, as the lag between electronic and mechanical hardware is present. The calculated number of steps does not give the accurate distance for the tool to travel. This is the reason one must develop a feedback mechanism which feeds the tool to the required and exact location by using a proximity sensor.

The chapter gives us a sense of how a G-code language takes its form using existing standards and a set sequence.

Along with this, the chapter demonstrates the layouts for the command identification, command formation, and command execution algorithms. It discusses the importance of these algorithms in developing an error-free program. These algorithms take us a step forward in writing the code for our microcontroller and giving life to our computer numeric control machine.

The analysis carried out using ANSYS clearly depicts all the various range of stresses, strains, and displacements upon which our design assemblies would be able to operate on.

Because of its flexibility in tasks and power over the axes for compound, complicated printing of metal components, it is known as the millennium of machining. To giving out a solid social effect upon the engineering sector, which not so far in the distant future will be utilizing IOT. The information associations will pave the way for the manufacturing of metal parts from work areas to complete houses. Besides, this section sums up the arrangement of wellbeing precautionary measures and wellbeing dangers related to a WAAM machine and what should be done so as to viably counter such risks.

As already stated, this was an industrial scale model of the 3-axis WAAM Machine, scaled up from a prototype. The WAAM machine has uncountable applications in the industry. This will give a huge head start for future projects in this regard.

Design methodology was adopted using the findings of the literature review. Almost all the parts that were used in the machine were standard parts and available on the market. This excluded the prolonged procedure of manufacturing individual parts.

Motor was selected using the torques required for each axis. This torque was calculated using the concepts learned in the Design of Machine Elements course. Simultaneously, the software part was also being coded. The language used to code the Arduino UNO was C, and the coding of the 3-axis milling machine was done using the methodology learned in the course CAD/CAM.

In this way, all the three main steps, i.e., the design and analysis, electronics, and software part, were done.

4.20.2 FUTURE RECOMMENDATIONS

This machine design can be scaled up to design an industrial scale 3-axis WAAM machine which can perform heavy-duty metal printing of parts with intricacy and precision.

Furthermore, a machine always has room for improvement, which in our case is the IOT. Using a better microcontroller, the stepping of the machine could be further divided to achieve the demanded accuracy.

A business development plan can be created to sell this machine in the market because 3-axis WAAM machines are not widely available, despite their ability to manufacture very complex 3D shapes.

REFERENCES

[1] P. Sadasivam and M. Amirthalingam, "Design and fabrication of micro-plasma transferred wire arc additive manufacturing system," *CIRP J. Manuf. Sci. Technol.*, vol. 37, pp. 185–195, 2022, doi: 10.1016/j.cirpj.2022.01.014.

[2] S. R. Singh and P. Khanna, "Wire arc additive manufacturing (WAAM): A new process to shape engineering materials," *Mater. Today Proc.*, vol. 44, no. xxxx, pp. 118–128, 2021, doi: 10.1016/j.matpr.2020.08.030.

[3] H. Wahid, H. Hashmi, M. U. Baig, N. A. Raza, M. F. Sheikh, and M. Y. Bhutto, "Development of CNC-based automated soldering machine," p. 17, 2022, doi: 10.3390/engproc2022020017.

[4] Z. Liu *et al.*, "Additive manufacturing of metals: Microstructure evolution and multistage control," *J. Mater. Sci. Technol.*, vol. 100, pp. 224–236, 2022, doi: 10.1016/j.jmst.2021.06.011.

[5] P. Chen *et al.*, "Recent advances on high-performance polyaryletherketone materials for additive manufacturing," *Adv. Mater.*, vol. 2200750, pp. 1–26, 2022, doi: 10.1002/adma.202200750.

[6] S. W. Williams, F. Martina, A. C. Addison, J. Ding, G. Pardal, and P. Colegrove, "Wire + arc additive manufacturing," *Mater. Sci. Technol. (United Kingdom)*, vol. 32, no. 7, pp. 641–647, 2016, doi: 10.1179/1743284715Y.0000000073.

[7] D. Ding, Z. Pan, D. Cuiuri, and H. Li, "Wire-feed additive manufacturing of metal components: technologies, developments and future interests," *Int. J. Adv. Manuf. Technol.*, vol. 81, no. 1–4, pp. 465–481, 2015, doi: 10.1007/s00170-015-7077-3.

[8] O. Abdulhameed, A. Al-Ahmari, W. Ameen, and S. H. Mian, "Additive manufacturing: Challenges, trends, and applications," *Adv. Mech. Eng.*, vol. 11, no. 2, pp. 1–27, 2019, doi: 10.1177/1687814018822880.

[9] M. T. Richardson, V. Patterson, and A. Parchment, "Microcontroller based space vector pulse width modulation speed control of three-phase induction motor," *in SoutheastCon 2021*, 2021, pp. 1–6, doi: 10.1109/SoutheastCon45413.2021.9401922.

[10] S. Ibaraki and R. Okumura, "A machining test to evaluate thermal influence on the kinematics of a five-axis machine tool," *Int. J. Mach. Tools Manuf.*, vol. 163, pp. 1–27, 2021, doi: 10.1016/j.ijmachtools.2021.103702.

[11] K. Lysek, A. Gwiazda, and K. Herbus, "Application of CAM systems to simulate of a milling machine work," *IOP Conf. Ser. Mater. Sci. Eng.*, vol. 400, no. 4, 2018, doi: 10.1088/1757-899X/400/4/042037.

[12] F. Veiga, A. Suárez, E. Aldalur, I. Goenaga, and J. Amondarain, *Wire Arc Additive Manufacturing Process for Topologically Optimized Aeronautical Fixtures*. Mary Ann Liebert, Inc., Publishers, 2021. doi: 10.1089/3dp.2021.0008.

[13] M. Dharmawardhana, A. Ratnaweera, and G. Oancea, "Step-nc compliant intelligent cnc milling machine with an open architecture controller," *Appl. Sci.*, vol. 11, no. 13, 2021, doi: 10.3390/app11136223.

[14] A. Oulasvirta, N. R. Dayama, M. Shiripour, M. John, and A. Karrenbauer, "Combinatorial optimization of graphical user interface designs," *Proc. IEEE*, vol. 108, no. 3, pp. 434–464, 2020, doi: 10.1109/JPROC.2020.2969687.

[15] M. Raji, A. Shokanbi, and H. Monday, "Design of ultra-low-end controllers for efficient stepper motor control," *MATEC Web Conf.*, vol. 160, pp. 1–5, 2018, doi: 10.1051/matecconf/201816002003.

[16] S. Chandrasekaran, S. Hari, and M. Amirthalingam, "Functionally graded materials for marine risers by additive manufacturing for high-temperature applications: Experimental investigations," *Structures*, vol. 35, no. January, pp. 931–938, 2022, doi: 10.1016/j.istruc.2021.12.004.

[17] M. Helu, A. Joseph, and T. Hedberg, "CIRP annals - Manufacturing technology A standards-based approach for linking as-planned to as-fabricated product data," *CIRP Ann. - Manuf. Technol.*, vol. 67, no. 1, pp. 487–490, 2018, [Online]. Available: 10.1016/j.cirp.2018.04.039

[18] K. Latif, Y. Yusof, and A. Z. A. Kadir, "Development of virtual component-based STEP-compliant CNC system," *Prog. Addit. Manuf.*, vol. 7, no. 1, pp. 77–85, 2022, doi: 10.1007/s40964-021-00215-0.

[19] G. M. Minquiz, V. Borja, M. López-Parra, A. C. Ramírez-Reivich, M. A. Domínguez, and A. Alcaide, "A comparative study of CNC part programming addressing energy consumption and productivity," *Procedia CIRP*, vol. 14, pp. 581–586, 2014, doi: 10.1016/j.procir.2014.03.009.

[20] M. Iqbal, A. Iqbal, M. M. Nauman, Q. Cheok, and E. Abas, "Manufacturing, remanufacturing, and surface repairing of various machine tool components using laser-assisted directed energy deposition," in *Functional Reverse Engineering of Machine Tools*, 2019, pp. 79–88. doi: 10.1201/9780429022876-7.

[21] B. Praveen, S. U. Abhishek, P. B. Shetty, J. Sudheer Reddy, and B. A. Praveena, "Industry 4.0 researchers computer numerical control machine tool to manufacture calligraphy board," *Lect. Notes Electr. Eng.*, vol. 790, no. December, pp. 197–206, 2022, doi: 10.1007/978-981-16-1342-5_15.

[22] M. Aamir, M. Waqas, M. Iqbal, M. Imran Hanif, and R. Muhammad, "Fuzzy logic approach for investigation of microstructure and mechanical properties of Sn96.5-Ag3.0-Cu0.5 lead free solder alloy," *Solder. Surf. Mt. Technol.*, vol. 29, no. 4, pp. 191–198, 2017, doi: 10.1108/SSMT-02-2017-0005.

[23] C. A. My, "Integration of CAM systems into multi-axes computerized numerical control machines," *Proc. - 2nd Int. Conf. Knowl. Syst. Eng. KSE 2010*, pp. 119–124, 2010, doi: 10.1109/KSE.2010.30.

[24] F. U. Khan and M. Iqbal, "Development of a testing rig for vibration and wind based energy harvesters," *J. Eng. Appl. Sci.*, vol. 35, no. 2, pp. 101–110, 2016.

[25] G. Singh *et al.*, "Experimental investigation and performance optimization during machining of hastelloy C-276 using green lubricants," *Materials (Basel)*, vol. 15, no. 15, p. 5451, 2022, doi: 10.3390/ma15155451.

5 Reverse Engineering of a 6 DOF Industrial Robotic Arm for Advanced Manufacturing Processes

Muftooh Ur Rehman Siddiqi, Riaz Muhammad, Yuchun Xu, Fatima Tu Zuhra, and Niaz Bahadur Khan

5.1 INTRODUCTION

Robots have drastically cut production time and costs, revolutionizing the manufacturing sector [1,2]. These engineering marvels now do tasks that are judged too risky or inappropriate for their human counterparts. It comes as no surprise that since 2012, the demand for robots in the manufacturing industry has increased by an astonishing 35% [3]. One particular type of robot, the robotic arm, is akin to a programmable mechanical limb with human-like functions [4–6]. These industrial robotic arms (IRAs) come in various axis configurations, with the majority featuring six-axis and being referred to as Six Degrees of Freedom (6 DOF) robots. With an almost spherical working envelope, the capabilities of these robots are nothing short of remarkable. They can move with incredible strength and speed, make intelligent decisions, and even exhibit self-sufficient and autonomous behavior [7].

Industrial robots, in contrast to humanoid robots, are machine tools that can have their actions changed by programming without modifying the hardware [4]. They combine the accuracy of a conventional machine tool with the adaptability of a human operator, giving them the best of both worlds. A robot can perform the same operation repeatedly with absolute accuracy, much like a machine tool, and it can also learn new tasks and employ different supplementary tools to broaden its range of physical abilities [4]. Four major stages can be used to summarize the development of the modern industrial robot.

DOI: 10.1201/9781003376620-8

5.1.1 STAGE 1: EMERGENCE

Modern industrial robots have their origins in the Renaissance, when Leonardo da Vinci developed concepts for robotic arms that eventually led to the development of humanoid robots like KAIST's Hubo in 2015. Several businesses have worked on developing a range of industrial robot technologies over many years. The first programmed robot arm was created in 1954 by George Devol and Joe Eagleburger, and in 1962, General Motors used it to operate the first industrial robot for performing risky and repetitive jobs on an assembly line [8]. The creation of the Rancho Arm in 1963 at the Rancho Los Amigos Hospital in California is another significant turning point in the history of industrial robotics [9]. In 1966, Kawasaki acquired from Unimation the right to produce industrial robot arms. In 1967, AMF Hermatool unveiled the Versatran industrial robot, which Japan imported. The Minsky-Bennet arm was developed in 1968 and was scientifically inspired by the claws and nerves of crayfish. Nachi robots were created in 1969 and are mostly used for handling parts and performing arc and spot welding [10]. General Motors was able to create 110 cars per hour due to significant automation in production systems, which inspired other companies to follow suit, including Fiat, BMW, Mercedes Benz, Volvo, and British Leyland.

5.1.2 STAGE 2: FUNDAMENTALS

The Stanford's arm was a six-axis articulated robot created in 1969 by Stanford University researcher Victor Scheinman that used all-electric technology to enable efficient and accurate movement [11]. The subject of robotics has since advanced because of this idea. Shakey, the first autonomous robot with mobility and artificial intelligence (AI), was presented a year later, in 1970 [12]. After this, the German company KUKA Robotics unveiled the Famulus, the first automated arm with six electric motors and six-axis mobility, in 1973 [13]. Fanuc introduced their robot devices that same year, which were mostly utilized for activities like drilling, molding, and wire cutting.

The robotics sector saw a tremendous growth surge in the 1980s, which was characterized by fierce competition and rivalry between businesses and their goods. The 1984 purchase of Unimation by Westinghouse Electric Corporation, a major robotics manufacturer, contributed to the industry's continued growth. However, due to consolidation brought about by increased competition, Stäubli Faverges SCA of France purchased Unimation in 1988. This business specialized in creating articulated robots for use in cleanroom settings among other industrial settings. Moreover, Stäubli Faverges SCA purchased Bosch's robotics division in late 2004 [11]. The 1989-founded Yaskawa Electric Corporation significantly contributed to the rise of multifunctional robots. Over 300,000 Yaskawa Motoman robots were in use worldwide as of 2021 [14].

5.1.3 STAGE 3: COMPUTING POWER

The time of technological growth in the computer industry saw a substantial transformation in the robotics sector. The introduction of the Motoman MRC in 1994, which enabled the synchronization of motion between two robots and had an exceptional capacity to manage 21 degrees at once, was one noteworthy advance [15]. The match in which IBM's Deep Blue robot upset Garry Kasparov, the reigning world chess champion, in May 1997 was a crucial turning point in the advancement of AI technology [15]. Another significant advancement occurred in 2000 when Honda completed the Advanced Step in Innovative-Mobility (ASIMO) after a decade of advancement. ASIMO represented a breakthrough in the design of humanoid robots and was equipped with advanced mobility and dexterity features. Furthermore, in 2014, NASA created the Integrated Structured Assembly of Advanced Composites (ISAAC), a large yet precise robot that was specifically designed for advanced composite manufacturing and assembly [15].

5.1.4 STAGE 4: INDUSTRY 4.0

The emergence of Industry 4.0 represents a convergence of various trends and technologies, potentially transforming manufacturing processes [3,16]. Industry 4.0, which includes a variety of technologies such as cyber-physical systems, the Internet of Things (IoT), artificial intelligence, and cloud computing, among others [17–21], entails the significant use of automation and data interchange in production settings. Manufacturers can run "Smart Factories" that can personalize products to specific consumer needs by utilizing Industry 4.0 technology [22–24]. This movement has been fueled by improvements in the transmission of digital instructions to the physical world and vice versa, such as advanced robots and 3D printing. Zero Down Time Solution (ZDT), a factory automation invention from General Motors, is one such example. GM created ZDT, a cloud-based software platform, in collaboration with FANUC and Cisco to evaluate data gathered from robots throughout their operations [25]. The global growth of the market for smart machines depends on the increasing acceptance of Industry 4.0 in sectors like automotive, electrical, and aerospace. The industry places a great emphasis on the predictable and dependable traits of industrial robots, as well as their accuracy and (relative) resistance to harsh environments [26].

Modern robots have weak cooperation with sensory input, despite having basic vision and touch capabilities [26]. Due to these drawbacks, robots are mostly utilized in highly structured industrial settings where the majority of variability and decision-making can be designed out of the work environment [27]. Industrial robots are used to carry out repetitive, preprogrammed activities including spot welding, spray painting, palletizing, and loading and unloading metal forming and cutting equipment [28]. However, the following generation of sensor-based robots will be more affordable and available, and they will be able to carry out a larger variety of activities under less regimented circumstances [29].

TABLE 5.1

A Comparative Analysis of Collaborative Robots

Manufacturer	Universal-Robots	KUKA-Robotics
Model	UR 10	LBR 14 R820
Maximum reach (m)	1.3	0.82
Payload (kg)	10	14
DOF (degree of freedom)	6	7
Weight of IRA (kg)	29.8	30

Future vision-enabled robots will be capable of carrying out operations including inspection, assembling, heat treatment, grinding, polishing, and electroplating [30]. As computer-controlled manufacturing systems replace manually controlled machines, several tasks achieved by production workers on the factory-floor will be accomplished by robots. Industrial robot sales have been increasing quickly, particularly in nations like the US, South Korea, China, and Germany [31]. The International Federation of Robotics claims that "Robot sales in 2014 climbed by 29% to 229,261 units, by far the largest annual total ever. Throughout time, sales of industrial robots to all industries rose" [32]. Prominent worldwide businesses are pioneering the development of cutting-edge robotic technology, including KUKA Robotics, Universal Robots, and Rethink Robotics. For instance, whereas KUKA Robotics offers a variety of SCARA configuration robotic arms, with its newest form, the KUKA LBR, competing with the UR10, Universal Robots only offers three main variants of robotic arms (the UR3, UR5, and UR10). The UR10 weighs about 30 kg, has a best reach of approximately 1 meter, and has six degrees of freedom. It can raise a burden of up to 10 kg (DOF). On the other hand, the KUKA LBR can lift a payload of up to 14 kg and has a maximum reach of 0.82 meters (7 DOF) [Table 5.1].

5.2 DESIGN AND ANALYSIS

Technical design specifications to reverse engineer the IRA are based on robotic arms available in the market, e.g., KUKA LBR and UR 10. Specifications are as follows: the robotic arm has a maximum reach of 1 meter, it is designed to handle a payload of 10 kg, has 6 DOF, and it can perform complex movements. Several conceptual models are designed based on the literature review. Every model has its pros and cons. For example, the joint-arm configuration conceptual model is for heavy payload, rigid, slow, and expensive while the second model has a higher DOF, is lightweight, requires no extra security measures, and is low cost. The conceptual designs of both models are shown in Figure 5.1. Figure 5.1a shows a simplistic but bulky 4 DOF design initially conceptualized with a lower number of components. As several issues are identified in this concept, multiple designs are conceptualized to solve those issues resulting in the final design shown in Figure 5.1, with a complex gear system.

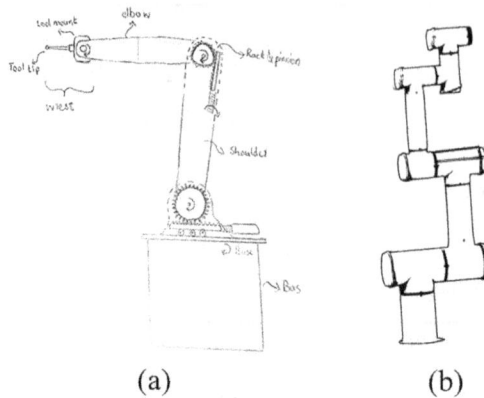

(a)   (b)

FIGURE 5.1   Conceptual models: (a) Initial design; (b) Final design.

The first iteration of the CAD model with the joint-arm configuration is shown in Figure 5.1. The initial model is large and intended for tasks that require significant strength. It operates at a slower speed due to its use of lead screw mechanism rather than haploid gears. It required an extra security measure for human workers in the industry. It has 4 DOF and the estimated cost of this model is high. After a series of iterations of conceptual designs based on torque calculations shown in Table 5.2 through Equation 5.1, a final design is selected and modeled as shown in Figure 5.2. Compared to previous models, this particular one is lighter in weight and operates at a higher speed. It has 6 DOF with which it can perform complex movements. This model does not require extra security measures. The estimated cost of this model is low and has presentable esthetics compared to the previous concepts. Under the previously mentioned specification, including the effect of body mass and payload, torque at various joints of the robotic arm is calculated through a mathematical equation.

$$T = F * R \tag{5.1}$$

where T = Torque (N-m), F is the force, and R is the moment arm.

TABLE 5.2

Torque of Joints

No. Links	Length of Link (cm)	Mass of Link (kg)	Torque at Joints (N-m)	Joints
1	9.22	5.2	11.6	Wrist joint 1
2	9.22	5.2	28	Wrist joint 2
3	11.57	5.2	54.6	Wrist joint 3
4	57.16	9.2	227.2	Elbow joint
5	61.27	12	477.2	Shoulder joint
6	18	9.4	550	Base joint

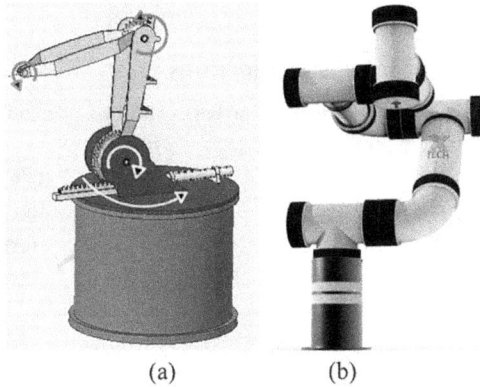

FIGURE 5.2   CAD model: (a) Initial design; (b) Final design.

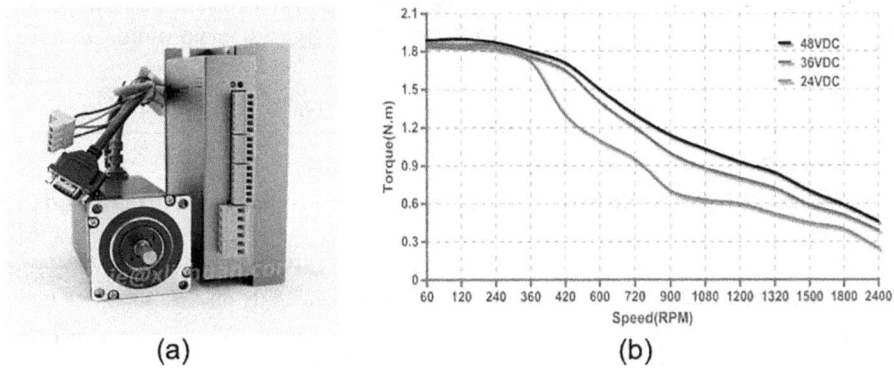

FIGURE 5.3   Lichuan hybrid servo motor and torque speed.

The minimum calculated torque is 11.6 N-m at the wrist joint-1. The maximum calculated torque is at the base joint which is 550 N-m [Table 5.2].

Selecting a motor with the right torque whilst keeping the weight of the structure low is targeted during calculations (Figure 5.3). To counter torques at different joints of the robotic arm, a variety of hybrid servo motors have been chosen. Hybrid servo motors are a more affordable and lightweight alternative to both AC and DC servo motors. Although hybrid servo motors occupy more space than other servo motors, they are still feasible to be used in robotics due to low prices. Each hybrid servo motor has an incremental encoder attached to it. The motors required for wrist joints are tabulated in Table 5.3. The hybrid servo motors are controlled by **LCDA86** and **LCDA808H** servo motion drives.

The maximum holding torque of the hybrid servo motor occurs at minimum speed and decreases as the motor speed increases. The cost and weight of a motor with the required torque are high and thus using it will increase the overall price of the robotic arm and load on the links. So, to increase the output torque of

TABLE 5.3

Lichuan's Hybrid Servo Motor Specifications

Motor Model	Step/ Angle	Length	Holding Torque	Current	Rotor Inertia	Encoder	Lead Wire	Weight
	°	mm	N-m	A	g.cm^2	PPR	No.	Kg
LC57H276	1.8	76+22	2	3	480	1000	4	1.2
LC86H2120	1.8	120+22	8.22	6	3600	1000	4	4
LC86H2160	1.8	156+22	12	7.5	5400	1000	4	5.4

motors, a gearbox is integrated for each hybrid servo motor. A comparison is drawn between a simple servo motor and a servo motor with a gearbox in Table 5.4. This configuration is shown in Figure 5.4, and it reduced the overall price and weight of the system. The selection of a gearbox reduction ratio greatly depends upon the holding torque and the speed of a hybrid servo motor as shown in Table 5.5.

TABLE 5.4

Comparison of Simple Servo Motor and the Servo Motor with Gearbox

Servo Motor		Servo Motor + Gearbox	
Torque Rated (N-m)	Price ($)	Torque Rated (N-m)	Price ($)
140	1850	150	480
250	2400	245	600

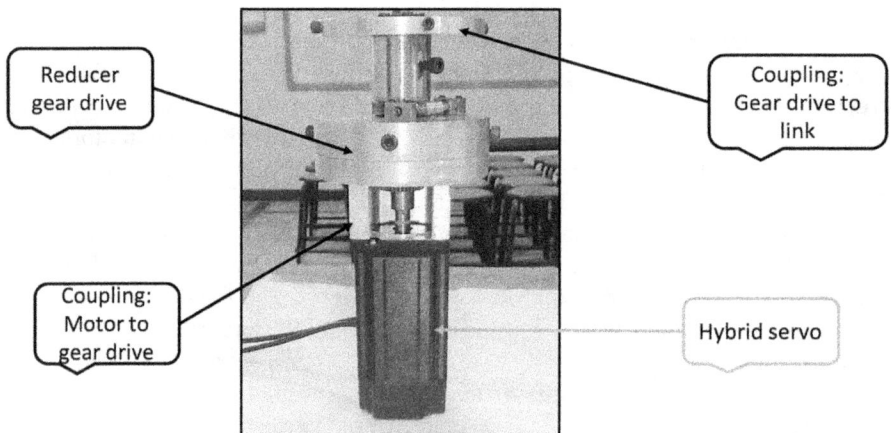

FIGURE 5.4   Basic assembly of motion component.

TABLE 5.5

Motor and Gear Selection

Joint	Applied Torque (N-m)	Motor Torque$_{Rated}$ (N-m)	Gear Ratio	Output Torque$_{max}$ (N-m)
Wrist joint-1	11.6	2	1:60	120
Wrist joint-2	28	2	1:50	100
Wrist joint-3	54.6	2	1:50	100
Elbow joint	227.2	8.2	1:50	410
Shoulder joint	477.2	12	1:50	600
Base joint	550	12	1:50	600

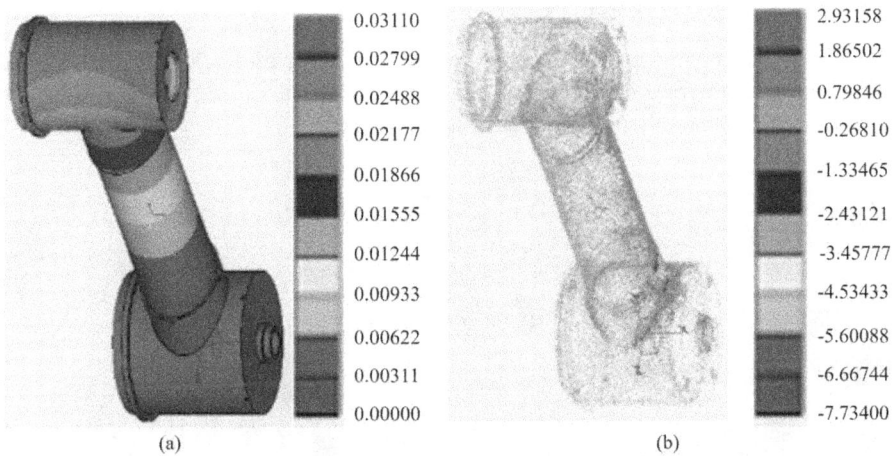

(a)	(b)
0.03110	2.93158
0.02799	1.86502
0.02488	0.79846
0.02177	-0.26810
0.01866	-1.33465
0.01555	-2.43121
0.01244	-3.45777
0.00933	-4.53433
0.00622	-5.60088
0.00311	-6.66744
0.00000	-7.73400

FIGURE 5.5   FEA results of shoulder joint: (a) Deformation; (b) Stress.

Finite Element Analysis for maximum load is carried out on the links of the final CAD model. One side of the link is fully fixed, while the load is applied at the other end. The applied load is based on the calculations provided in the tables previously. Failure criteria are based on strength of the two materials used in the analysis. The results are satisfactory with a maximum deformation of 0.03 mm shown in Figure 5.5. The material selected for the robot structure is 6061-T6 – 6000 Series Aluminum alloy based on FEA. Another FEA simulation is performed using uPVC, which confirms the material is safe to prototype IRA.

The hybrid servo motor is connected to its respective motion drive which controls its motion. The motion drive is controlled by the NVEM motion control card. The NVEM motion control card is controlled by Mac-3 software running on the computer having windows 7. The NVEM converts the information into signals (Pulse, Dir, Com, etc.) received from Mach3. The motion drive amplifies the signal to control the motion of the hybrid servo motor. This process is shown in Figure 5.6.

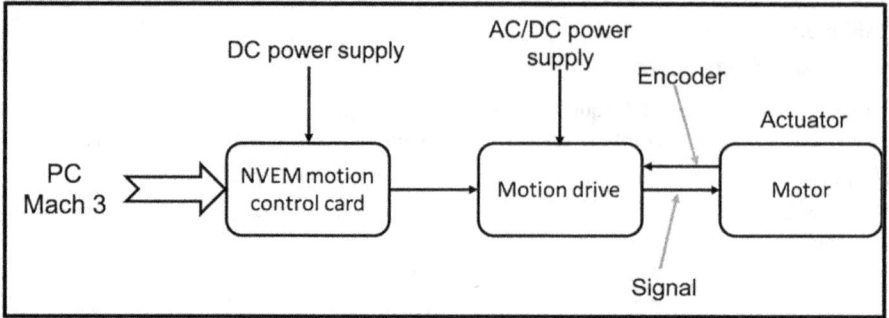

FIGURE 5.6 Flow chart of the control system.

5.3 MANUFACTURING AND TESTING

Gear drives are manufactured, heat treated, mounted for all six joints, and coupled to the respective hybrid servo motors as shown in Figure 5.7a. The Bearings, gears, shafts, and all the other components are assembled as shown in Figure 5.7b.

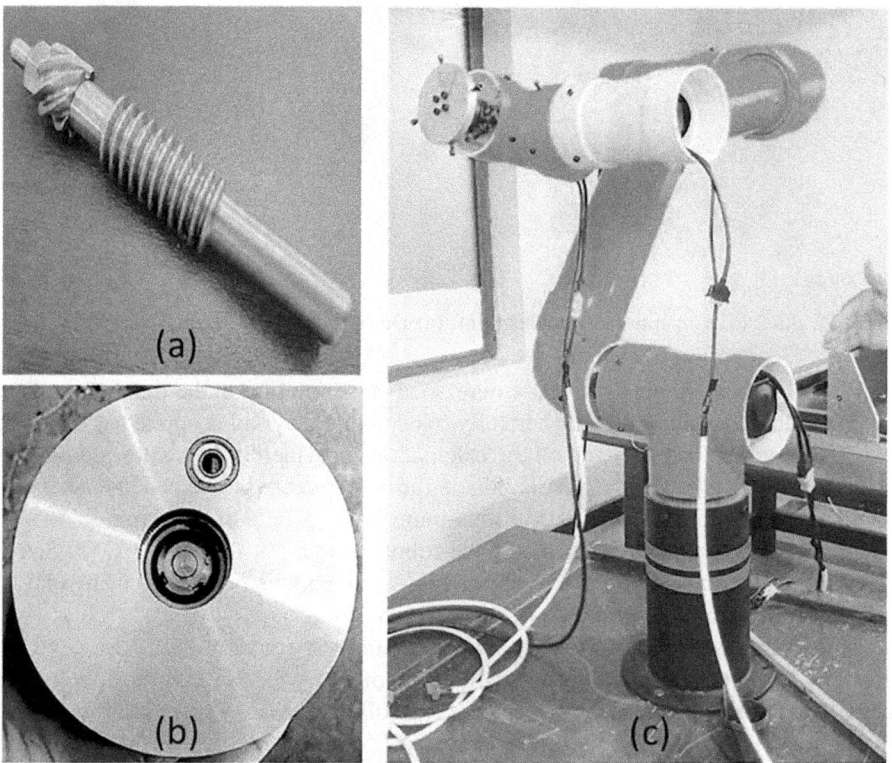

FIGURE 5.7 Manufacturing: (a) Worm and helical bevel gear; (b) Finished main plate of gear drive and bearing; (c) Prototype of IRA during assembly.

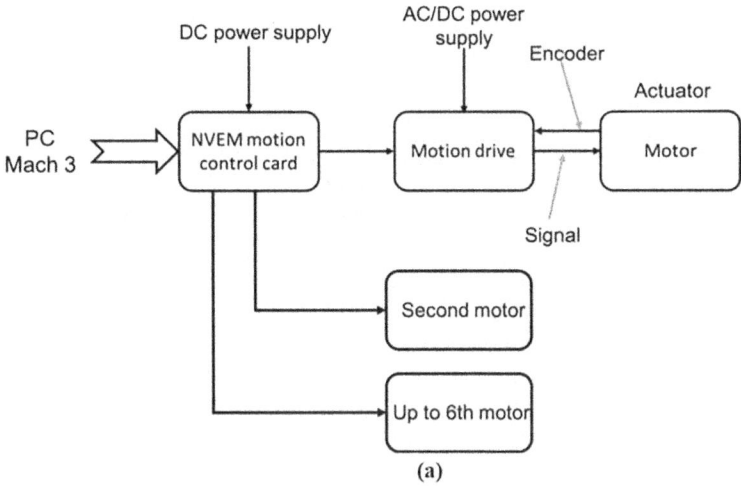

(a)

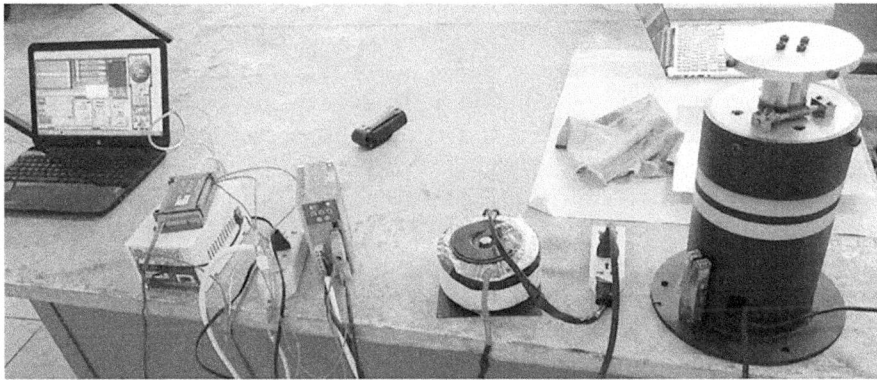

(b)

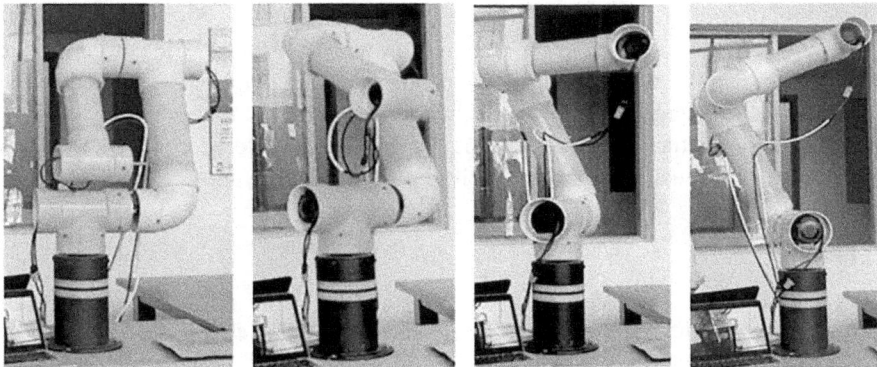

(c)

FIGURE 5.8   Motion component: (a) Logic; (b) Mechanism; (c) IRA's movements.

The motors are aligned to the gear drive via coupling. The links and joints of the prototype robot structure are fabricated from high-strength, low-weight Polyvinyl-chloride (PVC), and the base is manufactured with steel as shown in Figure 5.7c.

The reverse-engineered IRA has 6 DOF, and each joint has an electro-mechanical motion component, due to which a complex motion of the end effector is possible. Each of the electro-mechanical motion components consists of couplings (gear drive to link and motor to gear drive), reducer gear drive, and hybrid servo motor. Each motor has two cables, which are connected to the motor drives. The first cable is used by the motor drive to receive encoder signals to adjust motor movement and location. The second cable is used by the motor drive to send electrical current to run the motors as per the program. Cables are connected properly according to the respective manuals of NVEM, motor drives, hybrid servo motors, etc. The power cables are plugged into the power source. NVEM motion control card is accessed through Mach3 software through plug-ins placed in the library on a Windows 7 system. The NVEM converts the information into signals (Pulse, Dir, Com, etc.) received from Mach3. The motion drive then amplifies those signals to control the motion of the hybrid servo motor. The hybrid servo motors are closed loop and the position of the motor is sent back to the motor drive via an encoder cable. The flow chart of the process for hybrid servo motors controlled by the NVEM motion control card is shown in Figure 5.8a and its application is shown in Figure 5.8b. The 6 DOF movements of IRA are controlled for a pick and place operation and incremental sheet forming in the aerospace manufacturing industry as shown in Figure 5.8c.

5.4 CONCLUSION

A 6 DOF industrial robotic arm with over 1-meter reach and 10 kg payloads is reverse-engineered for advanced manufacturing processes. Sub-systems of the robotic arm, i.e., links, joints, mechanical power transmission, electrical power transmission, and software control elements are designed, produced, and tested. Sub-systems are integrated to produce the industrial robotic arm which is tested for advanced manufacturing processes. The robot is planned to be controlled through IoT and the cloud using an android cell phone in the next phase.

REFERENCES

[1] J. Schneider, D. Apfelbaum, D. Bagnell, and R. Simmons, "Learning opportunity costs in multi-robot market based planners," in *Robotics and Automation, 2005. ICRA 2005. Proceedings of the 2005 IEEE International Conference on*, 2005: IEEE, pp. 1151–1156.
[2] R. A. Bergs, *"Magnetically anchored 'reduced trocar' laparoscopy: Evolution of surgical robotics,"* Ms, The University of Texas at Arlington, 2006. [Online]. Available: http://pqdd.sinica.edu.tw/twdaoapp/servlet/advanced?query=1435992
[3] H. Lasi, P. Fettke, H.-G. Kemper, T. Feld, and M. Hoffmann, "Industry 4.0," *Business & Information Systems Engineering*, vol. 6, no. 4, pp. 239–242, 2014.

[4] R. Ayres and S. Miller, "The impacts of industrial robots," CARNEGIE-MELLON UNIV PITTSBURGH PA ROBOTICS INST, 1981.

[5] S. A. Siraj, "Critical analysis of press freedom in Pakistan," *Journal of Media and Communication Studies*, vol. 1, no. 3, p. 43, 2009.

[6] H. Rizvi, "The China-Pakistan Economic Corridor: Regional Cooperation and Socio-Economic Development. Vol. 34 and 35, Winter 2014 and Spring 2015, Numbers 4 and 1. Institute of Strategic Studies, Islamabad (ISSI)," ed, 2016.

[7] C. Breazeal and R. Brooks, "Robot emotion: A functional perspective," *Who Needs Emotions*, pp. 271–310, 2005.

[8] R. Ayres and S. Miller, "Industrial robots on the line," *The Journal of Epsilon Pi Tau*, vol. 8, no. 2, pp. 2–10, 1982.

[9] M. E. Moran, "Evolution of robotic arms," *Journal of Robotic Surgery*, vol. 1, no. 2, pp. 103–111, 2007.

[10] M. Wilson, *Implementation of robot systems: an introduction to robotics, automation, and successful systems integration in manufacturing*. Butterworth-Heinemann, 2014.

[11] M. Hägele, K. Nilsson, and J. N. Pires, "Industrial robotics," in *Springer handbook of robotics*. Springer, 2008, pp. 963–986.

[12] X. Chen, Y. Chen, and J. Chase, "Mobiles robots-past present and future," in *Mobile robots-state of the art in land, sea, air, and collaborative missions*. Intech, 2009.

[13] N. Jalil, *Material detecting and handling robotic arm*. Universiti Teknikal Malaysia Melaka, 2009.

[14] K. Okabayashi and K. Sakanashi, *Industrial robots controller*. Google Patents, 1999.

[15] H. Moravec, "When will computer hardware match the human brain," *Journal of Evolution and Technology*, vol. 1, no. 1, p. 10, 1998.

[16] V. Roblek, M. Meško, and A. Krapež, "A complex view of industry 4.0," *Sage Open*, vol. 6, no. 2, p. 2158244016653987, 2016.

[17] Y. Chen and H. Hu, "Internet of intelligent things and robot as a service," *Simulation Modelling Practice and Theory*, vol. 34, pp. 159–171, 2013.

[18] A. Junaid, M. U. R. Siddiqi, R. Mohammad, and M. U. Abbasi, "In-process measurement in manufacturing processes," in *Functional reverse engineering of machine tools*. CRC Press, 2019, pp. 105–134.

[19] M. U. Siddiqi *et al.*, "Low cost three-dimensional virtual model construction for remanufacturing industry," *Journal of Remanufacturing*, vol. 9, pp. 129–139, 2019.

[20] A. Junaid *et al.*, "Metrology process to produce high-value components and reduce waste for the fourth industrial revolution," *Sustainability*, vol. 14, no. 12, p. 7472, 2022.

[21] A. A. Riaz, R. Muhammad, N. Ullah, G. Hussain, M. Alkahtani, and W. Akram, "Fuzzy logic-based prediction of drilling-induced temperatures at varying cutting conditions along with analysis of chips morphology and burrs formation," *Metals*, vol. 11, no. 2, p. 277, 2021.

[22] N. Habib, M. Siddiqi, and R. Muhammad, "Thermal simulation of grain during selective laser melting process in 3D metal printing," *Journal of Engineering and Applied Science*, vol. 39, pp. 14–21, 2020.

[23] R. Muhammad, "A fuzzy logic model for the analysis of ultrasonic vibration assisted turning and conventional turning of Ti-based alloy," *Materials*, vol. 14, no. 21, p. 6572, 2021.

[24] M. Suleman, N. Nasir, M. U. R. Siddiqi, M. Usman, and S. Tariq, "3D printing of 2D nanomaterials," in *Smart 3D nanoprinting*. CRC Press, 2023, pp. 87–110.

[25] K. S. Low, W. N. N. Win, and M. J. Er, "Wireless sensor networks for industrial environments," in *Computational Intelligence for Modelling, Control and Automation, 2005 and International Conference on Intelligent Agents, Web Technologies and Internet Commerce, International Conference on*, 2005, vol. 2: IEEE, pp. 271–276.

[26] B. Singh, N. Sellappan, and P. Kumaradhas, "Evolution of industrial robots and their applications," *International Journal of Emerging Technology and Advanced Engineering*, vol. 3, no. 5, pp. 763–768, 2013.

[27] K. Gelli, *Sensor fusion for the intelligent control of a multilink robotic arm/hand system for target detection using neural networks*. Ms, Texas A&I University, 1993.

[28] Y. Wang, F. Zhang, and SpringerLink (Online service), *Trends in control and decision-making for human-robot collaboration systems*. Springer International Publishing: Imprint: Springer, 2017, pp. xix, 418 p. [Online]. Available: 10.1007/978-3-319-40533-9.

[29] M. I. Kalash, *Development and human factors analysis of pressure sensory substitution in robotic surgery*. Ms, Wayne State University, 2004. [Online]. Available: http://pqdd.sinica.edu.tw/twdaoapp/servlet/advanced?query=1419689

[30] P. Corke, and SpringerLink (Online service), *Robotics, vision and control: fundamental algorithms in MATLAB*, Second, completely revised, extended and updated edition. ed. (Springer tracts in advanced robotics, no. volume 118). pp. 1 online resource (xxix, 693 pages).

[31] J. Ray *et al.*, *China's industrial and military robotics development*. Defense Group, Incorporated, Center for Intelligence Research and Analysis, 2016.

[32] L. Fortunati, A. Esposito, and G. Lugano, *Introduction to the special issue "Beyond industrial robotics: Social robots entering public and domestic spheres*. Taylor & Francis, 2015.

6 Controller Design for a Six Degree of Freedom Parallel Robotic Machining Bed

Muhammad Faizan Shah, Zareena Kausar, and Muhammad Umer Farooq

6.1 INTRODUCTION

Machining is the process of removing material from a workpiece to achieve the desired shape and size [1]. High-cutting tool speeds are used in this process to remove material rapidly. Vibrations brought on by the rapid cutting speed result in errors in the surface finish and dimensions of the workpiece. When the tool moves in multiple directions, the inaccuracy increases. A solution proposed in [2] is to provide multiple degrees of freedom to the workpiece and confine tool motion to a single degree of freedom (DOF). The proposed mechanism is based on the parallel actuation configuration [1,3].

Recent studies explain that in comparison to serial manipulators, parallel mechanisms prevail favorable distinctiveness [4,5], such as high rigidity, stiffness, and load-carrying capacity [6–8]. Parallel Manipulators are prone to various types of errors as discussed in [9–16]. However, machining needs to be performed precisely and the consumers are concerned with the accuracy of the final product. For the proposed machining bed, errors due to the contemplating workpiece are studied in the previous work of the authors [17]. The error studied in [17], was studied as an open loop problem and no controller was suggested. As a first, this study proposes a conventional PID controller for tracking the trajectory of the workpiece placed on the moving top plate of the machining bed.

The control of six degrees of freedom parallel mechanism of the machining bed demands two significant problems to be taken into account. 1) Accurate positioning of the workpiece with respect to the cutting tool, and 2) Tracking of high-speed machining trajectory with speedy response. In the context of the former problem, literature presents studies. Wu et al. [18] presented an error model that analyzed the pose of the workpiece due to changes that occur in actuated and passive joints. Sun et al. [19] studied geometric errors using the screw theory. Zhan et al. [20] studied motion errors using analytical methods. Liu et al. studied geometrical errors

DOI: 10.1201/9781003376620-9

TABLE 6.1

Summary of Different Errors Studied by Researchers over the Years

Serial No.	Error Studied	Reference
1	Volumetric errors	[11]
2	Geometric errors	[14]
3	Position errors	[24]
4	Joint clearance errors	[13, 15]
5	Nongeometric errors	[25]
6	Pose errors by joint clearance	[10]

in low-mobility parallel manipulators [21]. Ni et al. [22] developed the position and orientation error models whereas the method for the reduction of structural and friction errors is studied by Shan and Cheng [23]. A summary of the various studies highlighting the errors that can occur in the parallel manipulators is presented in Table 6.1. All these studies tend to study the effect of these errors on the accuracy of the parallel manipulator.

Largely, PID control is used in the industries [26–28]. This is due to having a simple structure and robust operation [29]. For the proportional controller, the output is proportional to the error whereas the integrated controller integrates the value of error over time. Derivative controller depends on the rate of change of error. A PID controller is a combination of three controllers, combined to enhance the performance of the system by giving the desired output levels [30]. Based on its simple structure and convenience to implement experimentally, the PID controller is used in this study to minimize the trajectory errors of the proposed machining bed.

6.2 SYNTHESIS OF MACHINING BED KINEMATICS

Spatial motions of the moving top plate of the proposed machining bed about a fixed reference can be described using the kinematics model. The kinematic models for parallel manipulators have been widely discussed in the literature. Kinematic models are divided into two categories forward kinematics and inverse kinematics. Inverse kinematics refers to finding the values of the leg lengths whilst the orientation of the moving plate is known. Using the geometric parameters (Figure 6.1(b)), an inverse kinematics model was developed as in equation (6.1). In equation (6.1), Ln_i is the length of the linear actuator, T is the translation vector, UT_{Pi} is the distance between the center of the moving top plate and the anchor point where the linear actuator is mounted and UB_{Pi} is the distance between the center of the bottom fixed plate and the anchor point where the linear actuator is connected at the fixed plate. $_B^P R$ is the rotation matrix that relates to the motion of the moving platform in X, Y, and Z axes.

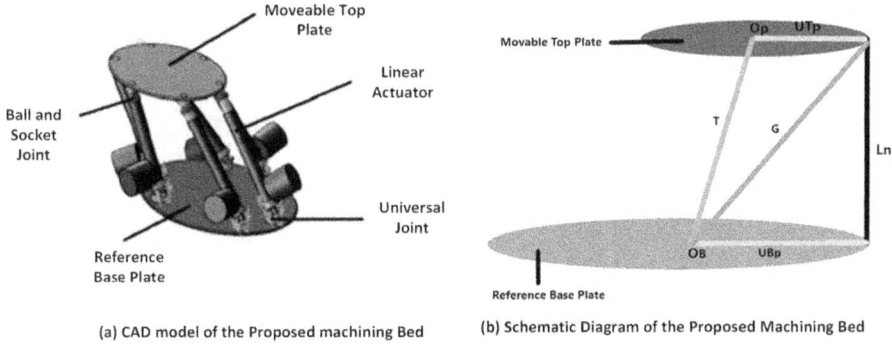

(a) CAD model of the Proposed machining Bed   (b) Schematic Diagram of the Proposed Machining Bed

FIGURE 6.1 Proposed parallel machining bed and its corresponding schematic diagram.

$$Ln_i = T + {}_B^P RUT_{Pi} - UB_{Pi} \qquad (6.1)$$

6.3 SYSTEM DYNAMICS

The dynamics model of the proposed parallel machining bed can be described as in equation (6.2). Where M is the system mass matrix, C is the matrix of Coriolis torques, and G is the vector of gravitational torques. In equation (6.2), T is the accumulative torque of the proposed parallel machining bed produced as a result of the forces applied.

$$M(\theta)\ddot{\theta} + C(\theta)\dot{\theta} + G(\theta) = T \qquad (6.2)$$

T can be calculated using relation (3). J is the Jacobian matrix that provides the mapping between task space torques and joint forces and is derived using the geometrical parameters of the machining bed presented in Figure 6.1(b).

$$T = J^{-1}F \qquad (6.3)$$

6.4 CONTROLLER DESIGN

As discussed earlier, PID is one of the most used controlling techniques in industrial robotics owing to its simplicity and easy implementation [20,28]. The control law for this technique is given by equation (6.4).

$$u(t) = K_p e(t) + K_i \int_0^t e(T)dt + K_d \frac{de}{dt} \qquad (6.4)$$

The controller is implemented by altering the gains K_p, K_i, and K_d. The schematic block diagram of the controller is shown in Figure 6.2. When used in tandem, the three control modes allow the controller to produce no steady-state error. The

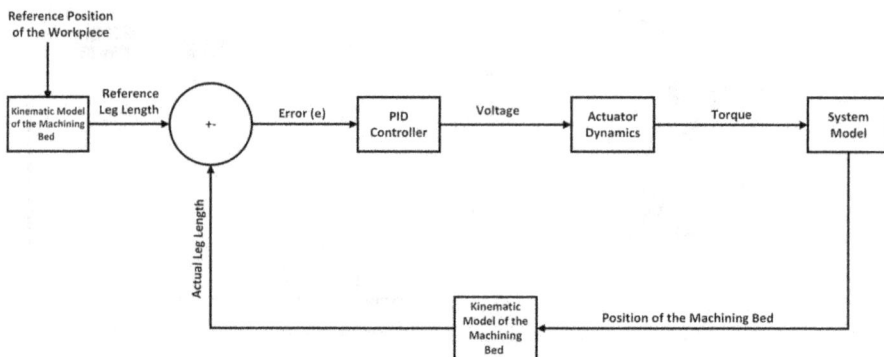

FIGURE 6.2 Schematic of the control algorithm followed for machining bed.

TABLE 6.2
Required Parameters for the Machining Bed

Parameters	Desired Range
Settling Time	>0.15 sec
Overshoot	>5%
Steady-State Error	>1%

proportionate parameter influences the existing errors. It curtails the rising time while reducing the steady-state inaccuracy. However, the steady error is not entirely eliminated. The integral parameter influences the response by influencing the accumulation of mistakes over time. The derivative parameter represents the error rate of change. Machining, as noted in section 6.1, is a procedure that demands tight tolerances on dimensions. When the desired accuracy of the workpiece is reached, the machining process is considered to be finished [30]. The motor speed of the linear actuator was intended to meet the requirements mentioned in Table 6.2 to obtain the minimal settling time.

For the machining bed to follow the given trajectory, it is necessary that parameters like settling time, overshoot, and steady-state error are in a certain range. The desired range for these parameters is presented in Table 6.2.

6.5 EXPERIMENTAL SETUP

The experimental setup of the proposed machining bed is shown in Figure 6.3. Six electric linear actuators with a maximum stroke length of 120 mm were used for actuation purposes in the machining bed. The proposed controller was implemented using the control scheme mentioned in Figure 6.2, by using an appropriate encoder. The controller was initially implemented to test the motion of the moving plate. For this purpose, a simple step input was applied, and the motion of the moving plate was observed. Further verification was carried out by

FIGURE 6.3   Experimental setup of the proposed machining bed.

TABLE 6.3
Test Description for the E-Shaped Workpiece to be Machined

Test No.	Points	Coordinates (mm)
1	1–2	[0;100;380]
2	2–3	[0;0;380]
3	3–4	[100;0;380]
4	4–5	No contact between the tool and workpiece
5	5–6	[200;100;380]
6	6–3	[200;0;380]

placing the E-shaped workpiece in the center of the moving plate and the machining operation was performed accordingly. Kinematic parameters for the E-shaped workpiece were obtained through the author's previous work presented in [2]. The test was divided into six parts as shown in Table 6.3.

6.6   RESULTS AND DISCUSSION

The experimental test was designed to implement the controller to minimize the error between the actual and desired trajectories. The desired trajectory for an E-shaped workpiece was obtained using the data presented in [2]. It was desired that the machining bed follows the desired trajectory with minimum overshoot and steady-state error, while having a settling time of less than 0.15 seconds.

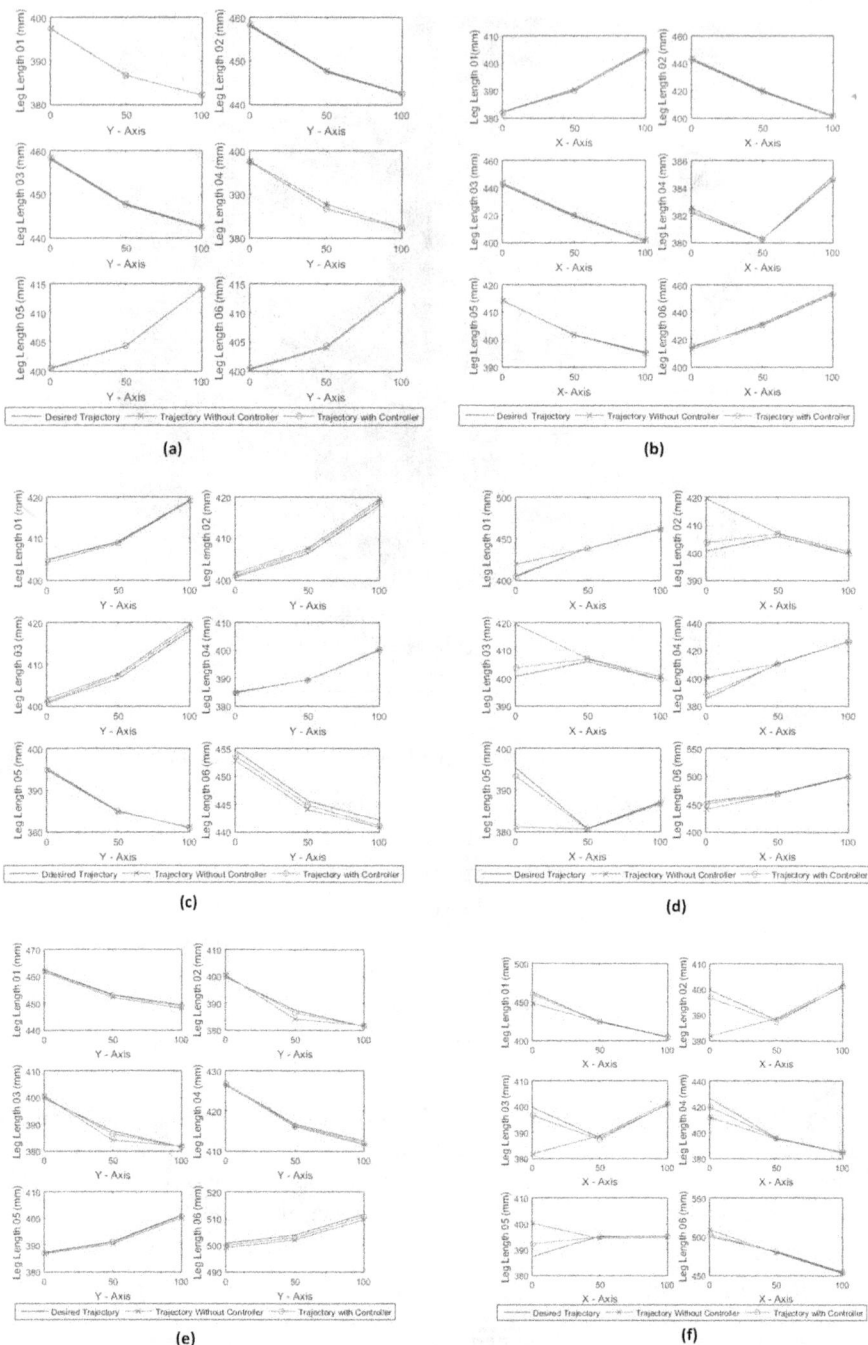

FIGURE 6.4 Various tests performed on the machining bed to verify the PID controller: (a) Test 01, (b) Test 02, (c) Test (3), (d) Test 04, (e) Test 05, and (f) Test 06.

Initially, minimal controller gains were provided. Although the machining bed worked fine a large error was observed in the actual and desired trajectories. For better accuracy, the controller gains were then tuned using the hit-and-trial method. Various tests were performed, and it was observed that the error kept on reducing with a gradual increase in the gains. However, when applied to a linear actuator, this increase in gains necessitates a higher torque, necessitating the use of a high-power source. This increase in torque affects the linear actuator's speed. The system has a rising time of 0.0167 seconds, a settling time of 0.0942 seconds, and a maximum peak amplitude of 1.13 at 0.0495 seconds. To minimize the overshoot of the overall system, the torque for the linear actuators was further increased. An overshoot of 2% was observed in the trajectory tracking of the machining bed which is less than the maximum overshoot limit set for the machining purpose. The results obtained for the six tests mentioned in Table 6.2 are shown in Figure 6.4. It can be seen that there is an undesired error between the actual and desired leg lengths when the controller is not implemented especially for tests 4, 5, and 6. This error is minimized with the help of the proposed controller. For tests 4 and 6, the error is reduced from the original but due limitation of torque applied by the linear actuator this is minimized to 3%. This error can be further reduced by designing an intelligent and optimal controller for the machining bed in the future.

6.7 CONCLUSION

This chapter discussed a six-degree-of-freedom machining bed for machining. The kinematic model of the machining bed, the error model of the machining bed, and the controller design to minimize error are the paper's key contributions. To practically examine the system, experimental experiments were carried out. The system attained the necessary leg lengths and error values, although at very high torque levels. An optimal controller will be built in the future for this purpose.

REFERENCES

[1] M. F. Shah, Z. Kausar, and F. K. Durrani, "Design, modeling and simulation of six degree of freedom machining bed," *Proc. Pakistan Academy Sci.*, vol. 53, no. 2, pp. 163–176, 2016.
[2] M. F. Shah, K. Nazeer, Z. Kausar, M. U. Farooq, S. S. Farooq, and G. M. Khan, "A six degree of freedom machining bed: Kinematic model development, verification, and validation," in *Functional Reverse Engineering of Strategic and Non-Strategic Machine Tools*, 1st ed., CRC Press, 2021, pp. 251–260.
[3] M. F. Shah, Z. Kausar, and M. U. Farooq, "Kinematic Modeling and Analysis of a 6 DOF Parallel Machining Bed," in *Proceedings - 22nd International Multitopic Conference, INMIC 2019*, 2019, pp. 1–5, doi: 10.1109/INMIC48123.2019.9022802.
[4] S. S. Farooq, A. A. Baqai, and M. F. Shah, "Optimal design of tricept parallel manipulator with particle swarm optimization using performance parameters," *J. Eng. Res.*, vol. 9, no. 2, pp. 278–295, 2021, doi: 10.36909/jer.v9i2.9073.
[5] Y. D. Patel and P. M. George, "Parallel manipulators applications—A survey," *Mod. Mech. Eng.*, vol. 02, no. 03, pp. 57–64, 2012, doi: 10.4236/mme.2012.23008.

[6] J. P. Merlet, "Jacobian, manipulability, condition number, and accuracy of parallel robots," *J. Mech. Des. Trans. ASME*, vol. 128, no. 1, pp. 199–206, Jan. 2006, doi: 10.1115/1.2121740.

[7] S. Pedrammehr, M. Mahboubkhah, and N. Khani, "A study on vibration of Stewart platform-based machine tool table," *Int. J. Adv. Manuf. Technol.*, vol. 65, no. 5, pp. 991–1007, 2013, doi: 10.1007/s00170-012-4234-9.

[8] Z. Yue, Y. Ye, and B. Gu, "Modeling and simulation of a 6-DOF parallel platform for telescope secondary mirror," *Ground-based Airborne Telescopes V*, vol. 9145, p. 91452P, 2014, doi: 10.1117/12.2055856.

[9] D. Sun, S. Hu, X. Shao, and C. Liu, "Global stability of a saturated nonlinear PID controller for robot manipulators," *IEEE Trans. Control Syst. Technol.*, vol. 17, no. 4, pp. 892–899, 2009, doi: 10.1109/TCST.2008.2011748.

[10] A. H. Chebbi, Z. Affi, and L. Romdhane, "Prediction of the pose errors produced by joints clearance for a 3-UPU parallel robot," *Mech. Mach. Theory*, vol. 44, no. 9, pp. 1768–1783, Sep. 2009, doi: 10.1016/j.mechmachtheory.2009.03.006.

[11] A. J. Patel and K. F. Ehmann, "Volumetric error analysis of a Stewart platform-based machine tool," *CIRP Ann. - Manuf. Technol.*, vol. 46, no. 1, pp. 287–290, Jan. 1997, doi: 10.1016/s0007-8506(07)60827-0.

[12] X. Li, X. Ding, and G. S. Chirikjian, "Analysis of angular-error uncertainty in planar multiple-loop structures with joint clearances," *Mech. Mach. Theory*, vol. 91, pp. 69–85, Sep. 2015, doi: 10.1016/j.mechmachtheory.2015.04.005.

[13] S. Erkaya, "Analysis of joint clearance effects on dynamics of six DOF robot manipulators," *Mech. Mach. Sci.*, vol. 24, pp. 307–314, 2015, doi: 10.1007/978-3-319-09411-3_33.

[14] H. Schwenke, W. Knapp, H. Haitjema, A. Weckenmann, R. Schmitt, and F. Delbressine, "Geometric error measurement and compensation of machines-An update," *CIRP Ann. - Manuf. Technol.*, vol. 57, no. 2, pp. 660–675, Jan. 2008, doi: 10.1016/j.cirp.2008.09.008.

[15] J. Meng, D. Zhang, and Z. Li, "Accuracy analysis of parallel manipulators with joint clearance," *J. Mech. Des. Trans. ASME*, vol. 131, no. 1, pp. 0110131–0110139, Jan. 2009, doi: 10.1115/1.3042150.

[16] K. L. Ting, J. Zhu, and D. Watkins, "Effects of joint clearance on position and orientation deviation of linkages and manipulators," *Mech. Mach. Theory*, vol. 35, no. 3, pp. 391–401, Mar. 2000, doi: 10.1016/S0094-114X(99)00019-1.

[17] M. F. Shah, Z. Kausar, M. U. Farooq, L. A. Khan, and S. S. Farooq, "Accuracy analysis of machining trajectory contemplating workpiece dislocation on a six degree of freedom machining bed," *Proc. Inst. Mech. Eng. Part C J. Mech. Eng. Sci.*, 2020, doi: 10.1177/0954406220974049.

[18] G. Wu, B. Shaoping, K. Jørgen A., and S. Caro, "Error modelling and experimental validation of a planar 3-PPR parallel manipulator with joint clearances to cite this version: HAL Id: hal-00832640 error modelling and experimental validation of a planar 3-PPR parallel manipulator with joint clearances," *J. Mech. Robot. Am. Soc. Mech. Eng.*, vol. 4, no. 4, 2012, Art. no. 041008.

[19] T. Sun, Y. Zhai, Y. Song, and J. Zhang, "Kinematic calibration of a 3-DoF rotational parallel manipulator using laser tracker," *Robot. Comput. Integr. Manuf.*, vol. 41, no. October 2016, pp. 78–91, 2016, doi: 10.1016/j.rcim.2016.02.008.

[20] Z. Zhan, X. Zhang, Z. Jian, and H. Zhang, "Error modelling and motion reliability analysis of a planar parallel manipulator with multiple uncertainties," *Mech. Mach. Theory*, vol. 124, pp. 55–72, 2018, doi: 10.1016/j.mechmachtheory.2018.02.005.

[21] H. Liu, T. Huang, and D. G. Chetwynd, "A general approach for geometric error modeling of lower mobility parallel manipulators," *J. Mech. Robot.*, vol. 3, no. 2, 2011, doi: 10.1115/1.4003845.

[22] Y. Ni, C. Shao, B. Zhang, and W. Guo, "Error modeling and tolerance design of a parallel manipulator with full-circle rotation," *Adv. Mech. Eng.*, vol. 8, no. 5, pp. 1–16, 2016, doi: 10.1177/1687814016649300.

[23] X. Shan and G. Cheng, "Structural error and friction compensation control of a 2(3PUS + S) parallel manipulator," *Mech. Mach. Theory*, vol. 124, pp. 92–103, 2018, doi: 10.1016/j.mechmachtheory.2018.02.004.

[24] A. Angelidis and G. C. Vosniakos, "Prediction and compensation of relative position error along industrial robot end-effector paths," *Int. J. Precis. Eng. Manuf.*, vol. 15, no. 1, pp. 63–73, Jan. 2014, doi: 10.1007/s12541-013-0306-5.

[25] C. Gong, J. Yuan, and J. Ni, "Nongeometric error identification and compensation for robotic system by inverse calibration," *Int. J. Mach. Tools Manuf.*, vol. 40, no. 14, pp. 2119–2137, Nov. 2000, doi: 10.1016/S0890-6955(00)00023-7.

[26] K. J. A. and T. Hägglund, *PID Controllers: Theory, Design, and Tuning, Second Edition*. ISA-The Instrumentation, Systems, and Automation Society, 1995.

[27] A. R. D. Tipi and S. A. Mortazavi, "A New Adaptive Method (AF-PID) Presentation with Implementation in the Automatic Welding Robot," in *2008 IEEE/ASME International Conference on Mechatronic and Embedded Systems and Applications*, 2008, pp. 25–30, doi: 10.1109/MESA.2008.4735715.

[28] M. A. Khosravi and H. D. Taghirad, "Robust PID control of fully-constrained cable driven parallel robots," *Mechatronics*, vol. 24, no. 2, pp. 87–97, 2014, doi: 10.1016/j.mechatronics.2013.12.001.

[29] E. M. Jafarov, M. N. A. Parlakçi, and Y. Istefanopulos, "A new variable structure PID-controller design for robot manipulators," *IEEE Trans. Control Syst. Technol.*, vol. 13, no. 1, pp. 122–130, 2005, doi: 10.1109/TCST.2004.838558.

[30] Z. Kausar *et al.*, "Energy efficient parallel configuration based six degree of freedom machining bed," *Energies*, vol. 14, no. 9, p. 2642, 2021.

7 Surface Normal Trajectory Generation for Robotic-Based Machining Operation on Curved Surface

Anton. R. Ahmad, Said G. Khan, Syed H. Shah, and Chyi-Yeu Lin

7.1 INTRODUCTION

Robotic-based machining for processes such as polishing and deburring has attracted many researchers for the last twenty years or more. Such algorithms can deal with a variety of machining operations [1,2]. However, most of the manufacturers either rely on skilled technicians or some expensive CNC machines may be able to deal with the complicated surfaces and shapes. These manual operations by skilled workers are usually very costly and time-consuming to maintain the required level of accuracy. On the other hand, robotic-based machining may be able to provide solutions to diverse machining problems [3–5].

A robotic machining cutting tool is mounted on the robot end-effector (see Figure 7.1).

Usually, radially available industrial robots are employed for robotic machining. One of the main requirements for quality machining is to maintain the contact force while tracking a surface. In manual machining, this will not be very easy. Therefore, there is a need for more novel and efficient constant force tracking strategies for robotic-based automated machining processes such as grinding and polishing of irregular and complicated geometries [6,7].

Alternate solutions such as CNC machines may be able to deal with the machining of different kinds of surfaces and produce highly accurate results [8]. However, CNC machines are very expensive and with a smaller number of functions and capabilities. In addition, the workspace size is usually limited and hence may not be able to deal with large workpieces such as turbine blades, etc. [9].

In order to provide an alternative and more versatile option for machining, the field of robotic-based machining is fast emerging. Many researchers around the world have proposed different control schemes for six degrees of freedom industrial

98

DOI: 10.1201/9781003376620-10

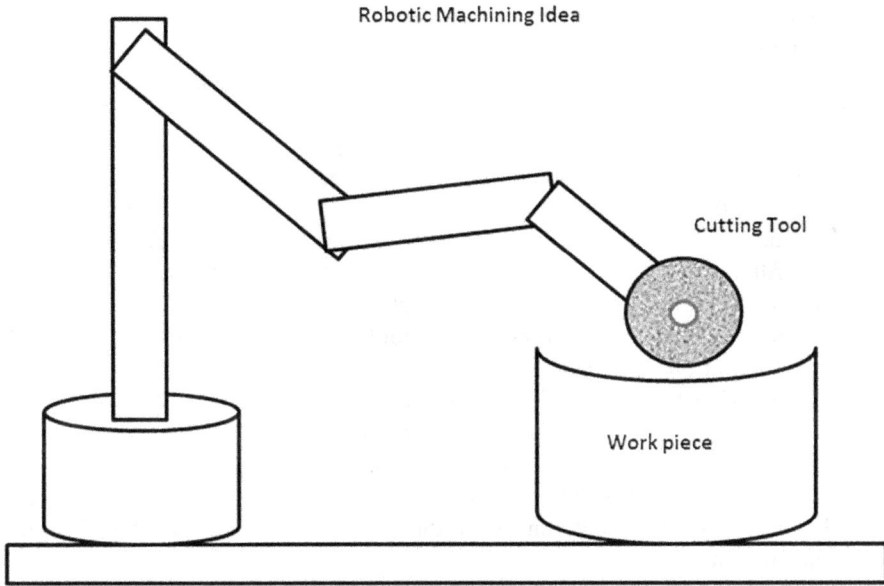

FIGURE 7.1   Robotic machining schematic.

robots employed for robotic machining. Suh schemes have significance in advancing the field [10,11]. However, still there are many unsolved problems such as dealing with curved surfaces. Furthermore, most of the research work only focuses on some aspects of robotics-based machining. Overall, robotic-based machining provides favorable surface finish quality, time efficiency, cost reduction, and convenient adaptability for repetitive processes [12]. For interested readers, the literature on kinematic redundant manipulators can be seen in [13].

As noted earlier, dealing with curved surface machining still poses challenges. Maintaining uniform force during robotic-based machining of irregular and curved surfaces requires new schemes to deal with this issue effectively.

The research in the area of robotic machining has great potential for industrial-scale machining processes. Therefore, this area has attracted many researchers to explore various possibilities to correctly position and orient the cutting tool connected to the end-effector of a robot manipulator for efficient and uniform cutting throughout the workpiece. Some of the most common machining processes such as deburring, grinding, and polishing were investigated [14,15]. Some recent work related to end-effector contact and non-contact trajectory tracking can be found in [16–18].

One of the main deficiencies in the previously cited work is that surface knowledge of the workpiece has been neglected. Knowledge of surface geometry is very essential for good-quality polishing. The contact force during polishing needs to be maintained 90 degrees to the surface to be polished. The work by the authors of [19] has noted that not only for the stability of the cutting it is essential to have the knowledge of the surface but also for the constant force maintenance during machining is extremely important. 3D models and coordinate measuring machines can also be employed for getting information on the surface to be machined. The drawbacks of such techniques make their technique less efficient: for instance, the uncertainty in the derived profile of the surface of interest.

In many cases, the orientation of the tool is disregarded and presumed to be fixed or derived from a computer-aided design (CAD) model utilizing modeling software to replicate the original surface, which poses a significant challenge. Consequently, obtaining and processing precise geometric information of a complex surface becomes arduous due to the robot's absolute and positioning error.

In this chapter, a novel robot end-effector force feedback-based technique for the position and orientation of the tool holder for real-time trajectory tracking aimed at curved surface machining is presented. The scheme employs a compliant passive tool holder driven by an algorithm to make the tool holder (cutting tool) normal to the surface and to maintain a constant force during cutting. The force feedback is provided by a six-axis force sensor mounted near the end-effector. Experimental results produced while tracking curved surfaces in real-time show the effectiveness of the proposed scheme and the designed compliant tool holder.

In this chapter, the main contributions of the work are the inherently flexible and low-stiffness tool holder, and active compliance control employing the force feedback from the force sensor. The experimental results prove the feasibility of the suggested approach.

The remainder of this chapter has been arranged in the following way. Following the Introduction, Contact-based Surface Estimation is presented. Afterward, the robot pose correction algorithm has been described. Then, simulation and experimental results are presented. Finally, the Chapter is summarized and concluded.

7.2   CONTACT-BASED SURFACE ESTIMATION

In this section, the concept of contact-based surface estimation is presented in detail. The proposed system of contact-based surface estimation needs a minimum of two fundamental components, such as a compliant contact tool and a multi-axis force/torque sensor. When the compliant contact tool makes contact with the surface, the force sensor detects the contact information and feeds it to the angle correction algorithm. The free body diagram (FBD) of the system is demonstrated in Figure 7.2, employing two-dimensional in order to simplify the explanation. The compliant contact pin is acting as a passive spring, attached to the multi-axis F/T sensor. The compliant contact pin is aligned with the z-axis of the multi-axis F/T sensor. When the compliant contact pin is in contact with the surface, the spring will exert some force on the surface. The multi-axis F/T sensor will measure the reaction force of the contact.

The resultant reaction force from the workpiece is perpendicular to the surface, which is commonly named Normal. The normal force is the result of three forces as mentioned in this Equation (7.1):

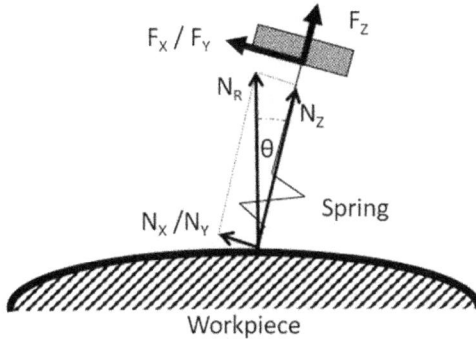

FIGURE 7.2   Free body diagram of the system.

$$\overrightarrow{N_R} = \overrightarrow{N_X} + \overrightarrow{N_Y} + \overrightarrow{N_Z} \tag{7.1}$$

where N_R represents normal force, N_X, N_Y, and N_Z are normal force components in Cartesian coordinates. N_Z is equal to the force that is exerted by the contact tool. When the compliant contact pin is aligned with the Normal, the reaction force is only equal to the force that is exerted from the spring, which is N_Z. When the compliant contact tool is not aligned with the Normal, other components of normal force, N_X and N_Y, will be non-zero. The angle of deviation from the normal can be drawn from the relation of the three forces as shown in Equations (7.2) and (7.3):

$$\overrightarrow{\theta_X} = \tan^{-1}\left(\frac{\overrightarrow{N_Y}}{\overrightarrow{N_Z}}\right), \tag{7.2}$$

$$\overrightarrow{\theta_Y} = \tan^{-1}\left(\frac{\overrightarrow{N_X}}{\overrightarrow{N_Z}}\right), \tag{7.3}$$

where θ_X and θ_Y are the angles of deviation from normal to the surface.

The contact-based estimation is to measure how much is the distance between the end-effector and the contact surface. The idea is to keep constant contact as well as the end-effector perpendicular to the surface. The distance of the end-effector to the contacted spot is estimated by using Equation (7.4):

$$L_S = L_O + \frac{F_Z}{k}, \tag{7.4}$$

where LS is the contact length of the tool. L_O represents the original length of the tool. F_Z is the force reading on the force sensor that is equal to the force of the spring exerted during compression. The term k is the spring constant. The spring employed in the design has linear behavior.

The angle deviation from the normal of the contacted spot will be estimated by employing Equations (7.5) and (7.6):

$$\Delta\theta_X = \tan^{-1}\left(\frac{F_Y}{F_Z}\right), \tag{7.5}$$

$$\Delta\theta_Y = \tan^{-1}\left(\frac{F_X}{F_Z}\right), \tag{7.6}$$

where F_X, F_Y, and F_Z indicate the force information gathered from the force sensor. Moreover, $\Delta\theta_X$ and $\Delta\theta_Y$ denote the angle adjustment necessary to achieve a perpendicular alignment between the robot end-effector and the surface being tracked.

7.3   ALGORITHM FOR POSE CORRECTION

The algorithm for robot pose correction is established using the information provided by the tool. The measured length and orientation of the end-effector with the contact surface will be used to correct the trajectory so that the desired distance and perpendicular relationship of the end-effector with the surface are obtained.

A transformation matrix is employed to correct the trajectory based on tool coordinates with respect to the robot base coordinate. The transformation matrix is given in Equation (7.7):

$$\begin{bmatrix} \hat{X} \\ \hat{Y} \\ \hat{Z} \end{bmatrix} = R \times \begin{bmatrix} L_T \times \sin(R_y) \\ L_T \times \sin(R_x) \cdot \cos(R_y) \\ L_T \times \cos(R_x) \cdot \cos(R_y) \end{bmatrix} + \begin{bmatrix} x \\ y \\ z \end{bmatrix} + \begin{bmatrix} -L_s \times \sin(R_y) \\ L_s \times \sin(R_y) \cdot \cos(R_y) \\ L_s \times \cos(R_x) \cdot \cos(R_y) \end{bmatrix} \tag{7.7}$$

where $\widehat{X}, \widehat{Y}, \hat{Z}$ are the updated position of the end-effector, whereas R_x and R_y are the current angles of the end-effector. Additionally, the current Cartesian position of the end-effector are denoted by x, y, and z. The target distance between the end-effector to the surface is L_T. L_s is the current length of the passive compliant tool. The term R is the rotation matrix which is given in Equation (7.8):

$$R = \begin{bmatrix} \cos\Delta\theta_Y & \sin\Delta\theta_X \sin\Delta\theta_Y & \cos\Delta\theta_X \sin\Delta\theta_Y \\ 0 & \cos\Delta\theta_X & -\sin\Delta\theta_X \\ -\sin\Delta\theta_Y & \sin\Delta\theta_X \cos\Delta\theta_Y & \cos\Delta\theta_X \cos\Delta\theta_Y \end{bmatrix} \tag{7.8}$$

where $\Delta\theta_X$ and $\Delta\theta_Y$, shows the angle adjustment needed to make the end-effector perpendicular to the surface. The tool coordinate system is converted into the flange coordinate system of the robot using this transformation matrix. The overall transformation matrix is shown in Figure 7.3.

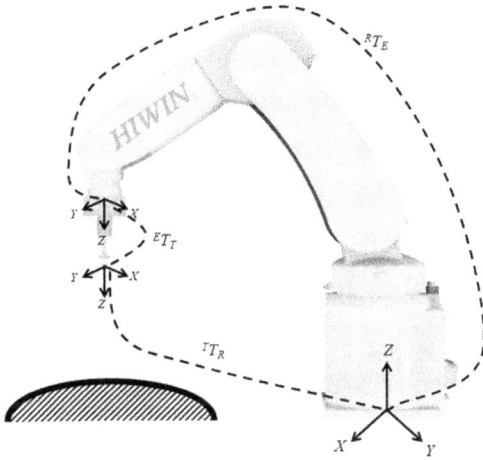

FIGURE 7.3 Overall transformation matrix.

7.4 SIMULATION RESULTS

The proposed system of contact-based surface estimation is initially tested via simulation. The simulation uses Autodesk Inventor, as shown in Figure 7.4. The compliant contact tool was facing a sloped surface of 30 deg. The spring used in the simulation has linear behavior. The tool pushes the spring so that the spring exerts a force on the surface. In the simulation, the compliant contact tool is moved down for several mm. It makes the spring compressed even more as shown in Figure 7.5b. In the simulation, the forces are monitored in the

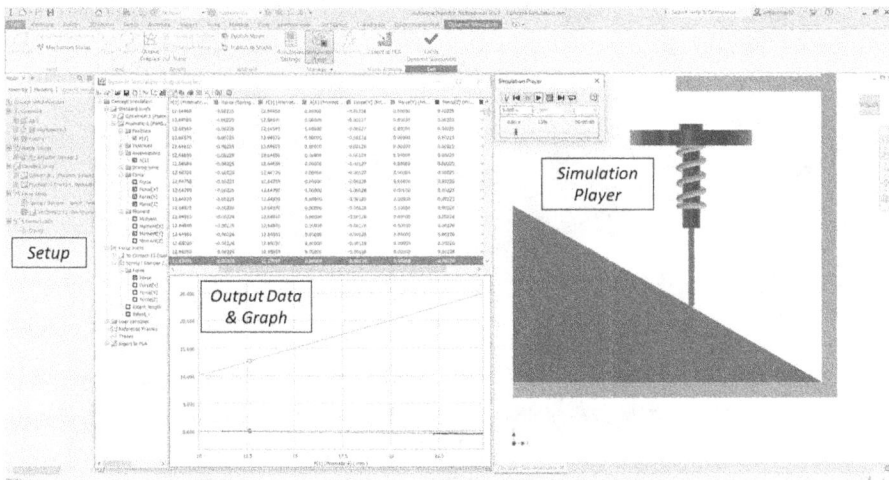

FIGURE 7.4 Simulation environment in Autodesk Inventor software.

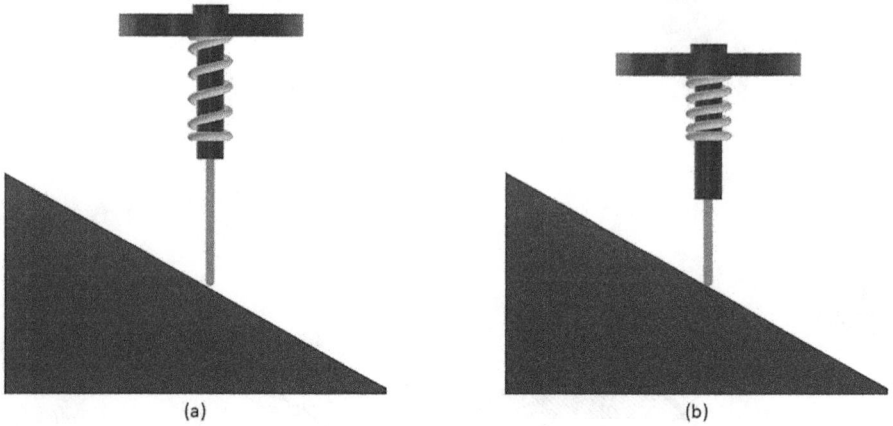

FIGURE 7.5 CAD dynamic simulation position (a) before contact and (b) after contact for 10 mm depth.

place of a multi-axis F/T sensor, which is on the base of the compliant contact tool. The forces that occur are FX and FZ since the slope occurs at the y-axis of the system. Since the nature of the spring is giving higher force when it's being compressed, as shown in Figure 7.6 simulation results, the forces of the base compliant contact tool are increasing linearly due to the movement of the tool head (ball).

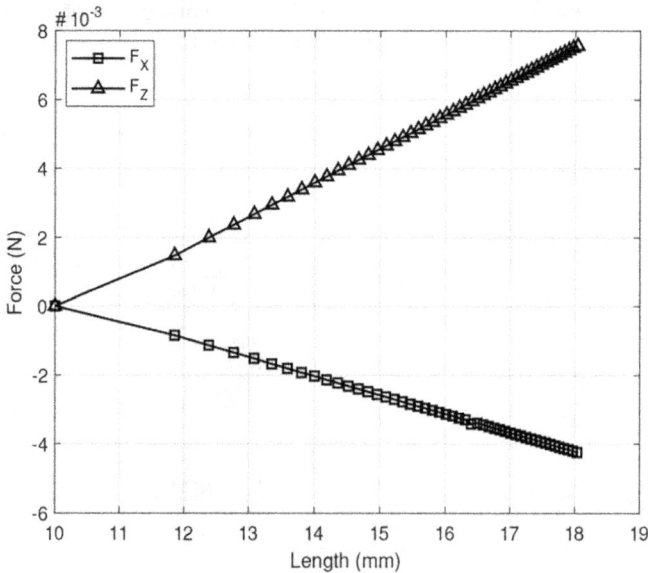

FIGURE 7.6 Simulation results at the base of the compliant contact tool.

TABLE 7.1
Angle Estimation Based on Force
Reading on Simulation

Parameters	$\Delta\theta_Y$
Average	29.3
Standard Deviation	0.1
Average Error	0.7

The force reading of the simulation is used to test the equation (7.6). As shown in Table 7.1, the average value of the angle Y, $\Delta\theta_Y$, is 29.3, which is 0.7 degree of error.

The simulation results validated the free body diagram and the equation (7.6) explained in the earlier Section.

7.5 EXPERIMENTAL RESULTS

In this section, the experimental setup for real-time implementation is briefly introduced and the experiments are carried out for validation. As mentioned earlier, a 6-axis force/torque sensor is employed for robot position and orientation adjustment. The compliant tool holder is mounted on a 6-DOF robotic manipulator (HIWIN), as shown in Figure 7.7.

The performance validation of the proposed system has been carried out employing three different experimental scenarios such as Customized surface, Flat surface, and Curved surface. The follow-up subsections present a detailed discussion of all the testing results obtained.

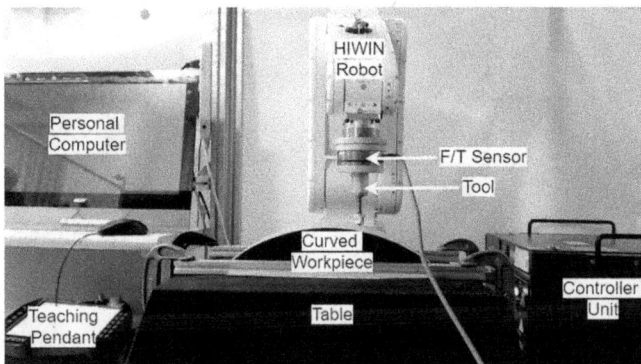

FIGURE 7.7 Experimental setup employed for the proposed system validation.

7.6 CUSTOMIZED SURFACE

Initially, the system performance was validated using a Customized 3D printed slope surface having a known surface angle offset from the normal such as 15°. The Cartesian coordinates of the robot on the surface with angle offset are known from the robot teaching pendant. Upon the tool's contact with the surface, the force sensor promptly captures and transmits the sensed forces to the algorithm executing on the computer. The algorithm computes the required angle to obtain perpendicular contact with the surface. This experiment was repeated several times to check the repeatability of the system. The system performance was significant on single point contact surface such as the average estimated angle was 14.3°, respectively on the customized workpiece with an average estimated error of 0.7°.

7.7 FLAT SURFACE

In the next step, the efficiency of the scheme was evaluated on the flat surface having a defined reference trajectory with random errors at each point. The reference trajectory with random error referred to as incorrect trajectory in the text was obtained manually by using a teaching pendant. The incorrect trajectory was given to the robot for surface tracking. The robot collects force information at each point of the incorrect trajectory and estimates the angle adjustment needed to obtain a normal trajectory for the surface. The obtained normal trajectory was compared with the desired normal trajectory. To simplify the discussion, the steps used for comparison on flat surface experiments are demonstrated in Figure 7.8.

The maximum error obtained in the angle after correction is −1.6°, while the maximum average depth's error was approximately −1.3 mm, respectively. The overall experimental results on flat surface are shown in Figures 7.9 and 7.10. The x-axis demonstrates the robot movement along the axis and the y-axis shows the angle in degree.

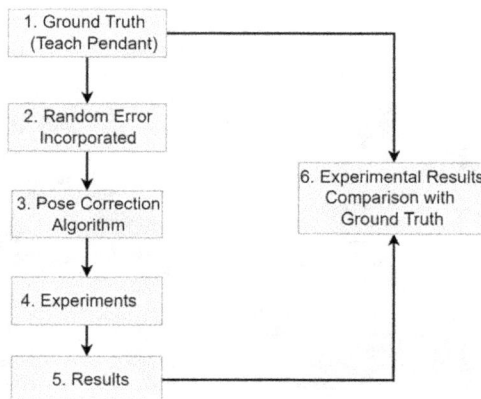

FIGURE 7.8 Steps to evaluate the efficiency of the scheme on flat surface.

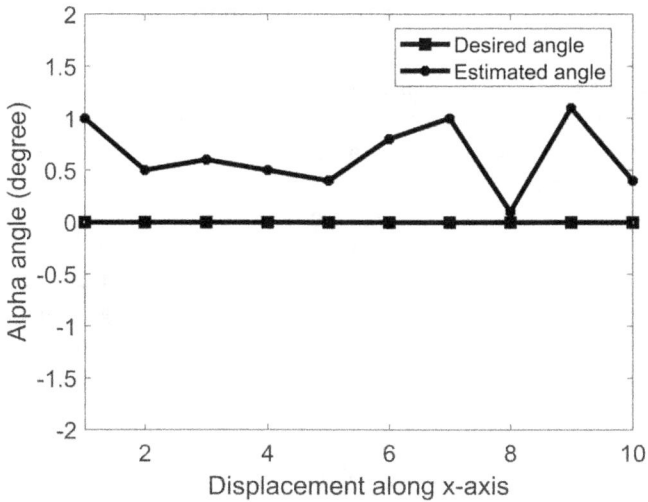

FIGURE 7.9 Experimental results on flat surface.

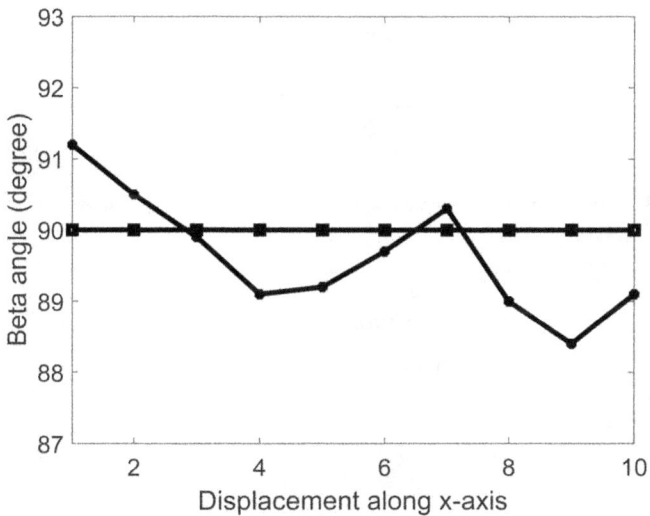

FIGURE 7.10 Experimental results on flat surface.

7.8 CURVED SURFACE

The curved surface experiment was conducted on a customized curved workpiece. The desired trajectory for this experiment was obtained from the CAD model of the workpiece. The reference trajectory (incorrect trajectory) was obtained manually using the robot teaching pendant. Unlike the flat surface reference trajectory, the curved surface reference trajectory was selected in such a way that the robot

Cartesian position was changed in all three x, y, and z directions while tracking the curved surface.

The robot was moving with constant speed and acceleration, the overall scenario of the curved workpiece tracking experiment as shown in Figure 7.11. After the robot tracked the reference trajectory a new modified trajectory was obtained as a result. The generated normal trajectory was compared with the desired normal trajectory obtained from the CAD model of the workpiece. The average maximum depth error is 1.23 mm while the maximum angle error noticed is 3.0°, respectively. The obtained result shows improvement to previously conducted studies [20,21]. A comparison graph of the desired angle and the estimated angle using the contact-based approach for the curved surface tracking is illustrated in Figure 7.12. This study shows promising potential for the automation of machining operations, such as robot laser cutting and grinding.

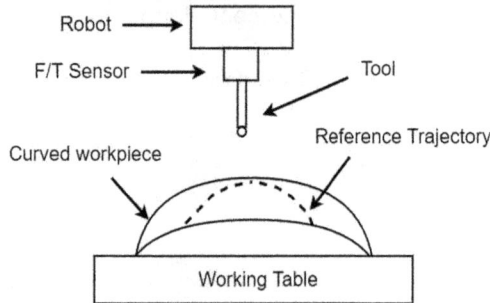

FIGURE 7.11   Scenario of the curved workpiece tracking experiment.

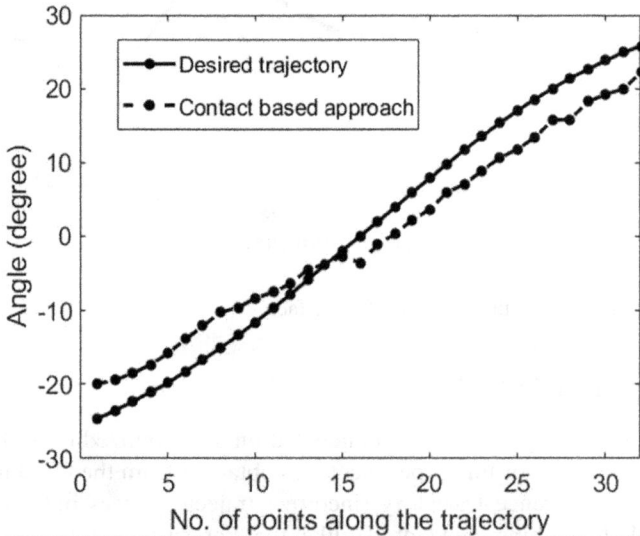

FIGURE 7.12   Comparison of desired and obtained normal trajectories.

7.9   CONCLUSION

In machining operations, irregular shapes and geometries require special attention from the machinists and usually require more time. Similarly, in robotic-assisted machining, dealing with curved surfaces (tracking trajectory) in real-time is even more challenging issue. In this chapter, we have presented a simple and more efficient solution for the robot tracking trajectory on curved surfaces aimed at machining operations such as debarring, grinding, and polishing. Our solution consists of a tool holder for the robot end-effector, a force sensor, and a simple program to implement a closed-loop scheme to adjust the tool holder's position and orientation. The six-axis force sensor has been installed at the robot end-effecter where the tool holder connects. In order to keep a constant force contact and to stay normal during tracking, the scheme automatically adjusts the z-axis position or the depth and the orientation of the tool i.e. angles with and x and y axes. Experimental results verified the effectiveness of the proposed algorithm. The average error in depth was recorded to be in the range of 0.2 to 1.2 mm, while the error in the orientation was found to be in the range of 0.5 to 3 degrees, respectively.

REFERENCES

[1] J. Li, T. Zhang, X. Liu, Y. Guan, and D. Wang, "A Survey of Robotic Polishing," *2018 IEEE Int. Conf. Robot. Biomimetics, ROBIO 2018*, no. December, pp. 2125–2132, 2018.

[2] I. Mohsin, K. He, Z. Li, and R. Du, "Path planning under force control in robotic polishing of the complex curved surfaces," *Appl. Sci.*, vol. 9, no. 24, 2019.

[3] A. E. K. Mohammad and D. Wang, "Electrochemical mechanical polishing technology: recent developments and future research and industrial needs," *Int. J. Adv. Manuf. Technol.*, vol. 86, no. 5–8, pp. 1909–1924, 2016.

[4] S. S. Martínez, J. G. Ortega, J. G. García, A. S. García, and E. E. Estévez, "An industrial vision system for surface quality inspection of transparent parts," *Int. J. Adv. Manuf. Technol.*, vol. 68, no. 5–8, pp. 1123–1136, 2013.

[5] J. Ernesto Solanes, L. Gracia, P. Muñoz-Benavent, J. Valls Miro, C. Perez-Vidal, and J. Tornero, "Robust hybrid position-force control for robotic surface polishing," *J. Manuf. Sci. Eng. Trans. ASME*, vol. 141, no. 1, pp. 1–14, 2019.

[6] M. J. Tsai, J. F. Huang, and W. L. Kao, "Robotic polishing of precision molds with uniform material removal control," *Int. J. Mach. Tools Manuf.*, vol. 49, no. 11, pp. 885–895, 2009.

[7] L. Liao, F. (Jeff) Xi, and K. Liu, "Modeling and control of automated polishing/deburring process using a dual-purpose compliant toolhead," *Int. J. Mach. Tools Manuf.*, vol. 48, no. 12–13, pp. 1454–1463, 2008.

[8] Y. Altintas, "Manufacturing Automation: Metal Cutting Mechanics, Machine Tool Vibrations, and CNC Design," *2nd ed. Cambridge: Cambridge University Press*, 2012.

[9] W. Ji and L. Wang, "Industrial robotic machining: a review," *Int. J. Adv. Manuf. Technol.*, vol. 103, no. 1–4, pp. 1239–1255, 2019.

[10] M. J. Tsai and J. F. Huang, "Efficient automatic polishing process with a new compliant abrasive tool," *Int. J. Adv. Manuf. Technol.*, vol. 30, no. 9–10, pp. 817–827, 2006.

[11] G. Wang, Q. Yu, T. Ren, X. Hua, and K. Chen, "Task planning for mobile painting manipulators based on manipulating space," *Assem. Autom.*, vol. 38, no. 1, pp. 57–66, 2018.

[12] C. H. Shih, Y. C. Lo, H. Y. Yang, and F. L. Lian, "Key ingredients for improving process quality at high-level cyber-physical robot grinding systems," *IEEE/ASME Int. Conf. Adv. Intell. Mechatronics, AIM*, vol. 2020–July, pp. 1184–1189, 2020.

[13] Y. S. Cheng, S. H. Yen, A. K. Bedaka, S. H. Shah, and C. Y. Lin, "Trajectory planning method with grinding compensation strategy for robotic propeller blade sharpening application," *J. Manuf. Process.*, vol. 86, pp. 294–310, Jan. 2023.

[14] S. H. Kim *et al.*, "Robotic Machining: A Review of Recent Progress," *Int. J. Precis. Eng. Manuf.*, vol. 20, no. 9, pp. 1629–1642, 2019.

[15] A. E. K. Mohammad, J. Hong, D. Wang, and Y. Guan, "Synergistic integrated design of an electrochemical mechanical polishing end-effector for robotic polishing applications," *Robot. Comput. Integr. Manuf.*, vol. 55, no. February 2018, pp. 65–75, 2019.

[16] C.-Y. Lin, C.-C. Tran, S. H. Shah, and A. R. Ahmad, "Real-time robot pose correction on curved surface employing 6-axis force/torque sensor," *IEEE Access*, vol. 10, no. August, pp. 90149–90162, 2022.

[17] A. R. Ahmad, C.-Y. Lin, S. H. Shah, and Y.-S. Cheng, "Design of a compliant robotic end-effector tool for normal contact estimation," *IEEE Sens. J.*, vol. 23, no. 2, pp. 1515–1526, 2023.

[18] S. H. Shah, C.-Y. Lin, C.-C. Tran, and A. R. Ahmad, "Robot pose estimation and normal trajectory generation on curved surface using an enhanced non-contact approach," *Sensors*, vol. 23, no. 8, 2023.

[19] P. R. Pagilla and B. Yu, "A stable transition controller for constrained robots," *IEEE/ASME Trans. Mechatronics*, vol. 6, no. 1, pp. 65–74, 2001.

[20] S. Wang, S. Chung, O. Khatib, and M. Cutkosky, "SupraPeds: Smart staff design and terrain characterization," *IEEE Int. Conf. Intell. Robot. Syst.*, vol. 2015–Decem, pp. 1520–1527, 2015.

[21] B. Wang, J. Li, H. Chen, Y. Guan, and T. Zhang, "A Normal Tracking Method for Workpieces with Free-Form Surface in Robotic Polishing," *Mech. Mach. Sci.*, vol. 111, no. March, pp. 1753–1765, 2022.

Part 4

4th Industrial Revolution

8 Applications of Digital Twins in the Development of a Predictive Maintenance System

Dan Noje and Radu Tarca

8.1 INTRODUCTION

In Industry 4.0 [1], process automation plays a crucial role [2]. Process automation is based on data collected and transmitted in near real time by various sensors, generically referred to as IoT devices [3]. Of the many processes that can be automated, the development of automated information systems for predictive maintenance (PrM) [4,5] is particularly important. The obvious benefits of predictive maintenance [6] in terms of reducing manufacturing costs [7,8], increasing the reliability of systems, and increasing the productivity and predictability of manufacturing processes [9,10], have made it an indispensable tool in the manufacturing sector and beyond.

An objective in developing manufacturing infrastructures is to use sustainable processes. One way to achieve this goal is the use of predictive maintenance [11], which leads to an efficient and seamless manufacturing process [9]. Current research results in development of low-cost and efficient IoT (Internet of Things) devices, impacted in a positive way by the development of PrM systems. Since the newly developed PrM systems are both reliable and cost effective, they start to be used in industries where it was not technically or economically possible [9]. PrM systems analyze the evolution of parameters that characterize industrial equipment or processes to trigger maintenance tasks only when required, preventing the occurrence of major failures [9].

IoT-enabled PrM systems use machine learning (ML) algorithms to analyze the acquired data and to compute the occurrence possibility of a malfunction or if equipment or part of its components is too worn [12]. In order to develop efficient PrM systems, the data used to feed the decision algorithms has to be accurate and reliable in terms of measurement and acquisition [13]. Among all the artificial intelligence algorithms and tools, artificial neural networks are generally preferred for their accuracy in data analysis [14,15]. The actual trend in developing PrM

DOI: 10.1201/9781003376620-12

systems is to combine artificial intelligence (AI) algorithms with data acquisition and processing in a virtual environment [16], called digital twin (DT) [17], to achieve the best results in terms of efficiency and cost-effectiveness. Taking advantage of research and innovation, a new industrial phase is reached, the so-called "Industry 5.0" [18] phase in which the focus is on the transition toward a sustainable, human-centered, and resilient European industry. This emphasizes the use of virtual environments and artificial intelligence for design, testing, and prototyping, thus reducing the time and constancy required for these steps before a working model is realized.

8.2 DIGITAL TWINS – BASIC CONCEPTS

Digital twins are bringing major transformations in terms of production, monitoring, and maintenance processes, leading to significant decreases in production costs, product development time, and innovation [17]. It should be noted that the development of IoT devices and technology, as well as artificial intelligence, has enabled the increasingly rapid development of industrial digital twin technology. Simplistically speaking, the digital replica of an entity is fed with data streams acquired from the real environment by IoT devices. This allows analysis of how this entity would function and respond in the event of an actual development. The data thus obtained is processed and analyzed, and based on it, new iterations of the development process can be decided without the need for a real prototype.

Digital Twin Consortium defines the concept of digital twin as "a virtual representation of real-world entities and processes, synchronized at a specified frequency and fidelity" [19].

To better understand why it is necessary to implement the digital twin concept, we will start from those characteristics that a company must implement in order to be considered a successful one. Those characteristics are presented in Table 8.1.

TABLE 8.1
Reasons to Implement Digital Twin Concept [20]

Speed	Agility	Data Driven Enterprise	Competitiveness
Necessity to be able to deliver new capabilities from the enterprise faster than ever	Strong focus on rapid response to the customer needs, while still controlling costs and quality	To develop a data driven enterprise to be able – in real time – to collect, integrate, and analyze data, to obtain a strategic advantage	Strengthening innovative capabilities and streamlining processes through an efficient management

TABLE 8.2
Types of Digital Twins

Digital Twins Product	Digital Twins Production	Digital Twins Performance
Digital twins that are used to develop new products.	Digital twins that are used to verify manufacturing processes.	Digital twins that are used to collect and analyze data, and to make decisions based on performed analyses.
The Digital Twin Product comprises the entire product, its hardware, software, mechanics, actuators, electrics, and physical behavior. This way is possible to simulate and validate each step of its development. More we have the possibility to identify problems and possible failures before producing real parts.	Using Digital Twin Production is possible to plan the entire manufacturing process in a virtual/augmented environment. Using those tools, it is possible to realize the plant layout, visualize the material/information flows and also to detect possible bottlenecks, to simulate the PLC code, and robot/machine tools program for the automation hardware, ultimately leading to a virtual commissioning that helps to test and optimize the virtual production lines in order to reduce time, effort, and risk for the real commissioning.	The Digital Twin Performance encompasses both product and production performance. It is continuously fed with data delivered by both product and production facilities, leading to new insights. Having the connection with integrated automation components, the manufacturing facility provides all relevant data, which is then analyzed with specialized software that enables continuous optimizations along the entire value chain.

There are usually three types of digital twins, as presented in Table 8.2.

The functional integration of the three digital twins: product, production, and performance generates the concept called digital thread.

Based on these characteristics of digital twins, it is easy to identify the offered benefits. These include [21]:

- engineers gain insight into the real-world use of the products they have designed, with all the benefits this brings;
- advanced ways of maintaining and managing products and assets are within reach, as there is a digital twin on which to test intervention without the danger of damaging the real one;
- decisions can be made in complex environments, as it is done in the aviation industry [22] or personalized medicine [23];
- another benefit of using digital twins is due to the connection to their real-world twins. It provides the possibility of developing both of them to deliver new performance, or to lead to the development of new services or products.

A proof of the efficiency and importance of using digital twin technology in the development process of new products, even of several versions in parallel, is the fact that this technology is currently used in the aerospace industry [22], and predominantly in the development of 6th generation fighter aircraft [24,25]. In this particular case, digital twin is also used to model the environment in which the new prototype aircraft is to evolve and how they interact.

Another extremely important area in the current economic context is predictive maintenance and how digital twin technology can be used to develop an automated system for predictive maintenance.

8.3   APPLICATIONS OF DIGITAL TWINS IN THE DEVELOPMENT OF A PREDICTIVE MAINTENANCE SYSTEM

Over the next few years, the level of enterprise digitization is expected to double, and predictive maintenance is expected to lead the investment race, with more than half of enterprises piloting predictive maintenance initiatives, some of which will benefit from digital twin technology as a key enabler [26]. Due to the increasing adoption of the use of predictive maintenance solutions using digital twin technology, supporting tools have emerged for prototyping these solutions, such tools are provided by the MathWorks platform [27].

Using digital twin technology, solutions can be implemented not only to determine the remaining useful life of equipment (RUL) but also to ensure predictive maintenance of equipment using health forecasting and management techniques [28]. According to the same sources, the stages of implementing such a system involve:

- modeling in a digital environment capable of simulating the behavior of the real machine of its resources and properties;
- data collection by machine controllers and IoT devices to be used for synchronous tuning of digital models and their simulated counterparts;
- simulation output is then used to assess the state of the resources and to calculate RUL. In this way, the condition and status of machines can be monitored and predicted as a result of simulation, without invasive techniques of common predictive maintenance solutions.

In order to have a complete structure of the desired maintenance system, the resulting solution must be integrated with dedicated software to manage maintenance tasks [29], and for the implementation to be reliable, the data collected by machine controllers and various IoT devices must be passed through a validation module [30] before they are used to assess the condition and state of the machine. In [30] the validation module was implemented using Shepard local approximation operators defined in Riesz MV-algebra [31,32], based on the Riesz MV-algebra structure of IoT devices signals, but any other approximation methods may be tested.

Possible applications of digital twin in the development of a predictive maintenance system include:

- simulations of disturbances in the system in which a given piece of equipment operates. We consider as part of these disturbances the environmental disturbances (e.g., vibrations, temperature, or humidity variations).
- simulations of the occurrence of faults or malfunctions in a piece of equipment.
- simulations of the triggering of alerts and predictive maintenance tasks in order to be able to track the speed of intervention and the elimination of simulated problems duration.

Sometimes a major impediment in the previously mentioned simulations is to feed the digital twin with appropriate data because we do not always have access to real data from controllers or IoT devices that correspond to such situations, or it would imply damaging some real equipment which would increase the development costs. This led us to the idea to consider the same approach as used in the development of new fighter aircrafts, meaning to use a digital twin or several digital twins to simulate the environment in which the machine operates. This environment represents the machine controllers and the set of IoT devices that provide the test data. To achieve the planned results, using Machine Learning techniques, the digital twin should be trained to generate the desired errors either one at a time or in simulations of two or more overlapped errors. This way, having control over which possible errors we test would be much faster and cheaper to get the desired results in developing the predictive maintenance system. Using this controlled way to generate the data feed for the predictive maintenance system, all three possible applications can be implemented and tested.

Another improvement that can be added to a predictive maintenance system is to transfer compressed data [33] from controllers and IoT devices to the predictive algorithms used by the predictive maintenance system. In [33] were proposed two options to implement this compression:

- to use a lower resolution of acquired parameters and later enhance it;
- to compress the acquired data, and after its transfer through the network to decompress it.

In [33] the first option was chosen. The test implementation was performed on digital thermal images acquired using a low-resolution temperature sensor. After the data was transferred, it was enhanced using an algorithm, which is actually an image zooming algorithm, based on Shepard local approximation operators.

To use this compression feature in a live predictive maintenance system, it is much easier, safer, and less expensive to test it first on a digital twin, and after the results are the desired ones to deploy it on production systems. The digital twin can test several compression/decompression algorithms and parametrizations without risking bad predictions due to improper data quality.

To develop a Digital Twin Product application, a Mechatronics Concept Designer (Simens [34]) was used and models of digital twins products were developed on some components of the UO-01-FMC flexible manufacturing cell [35–37] (Figure 8.1 and Figure 8.2) from the University of Oradea.

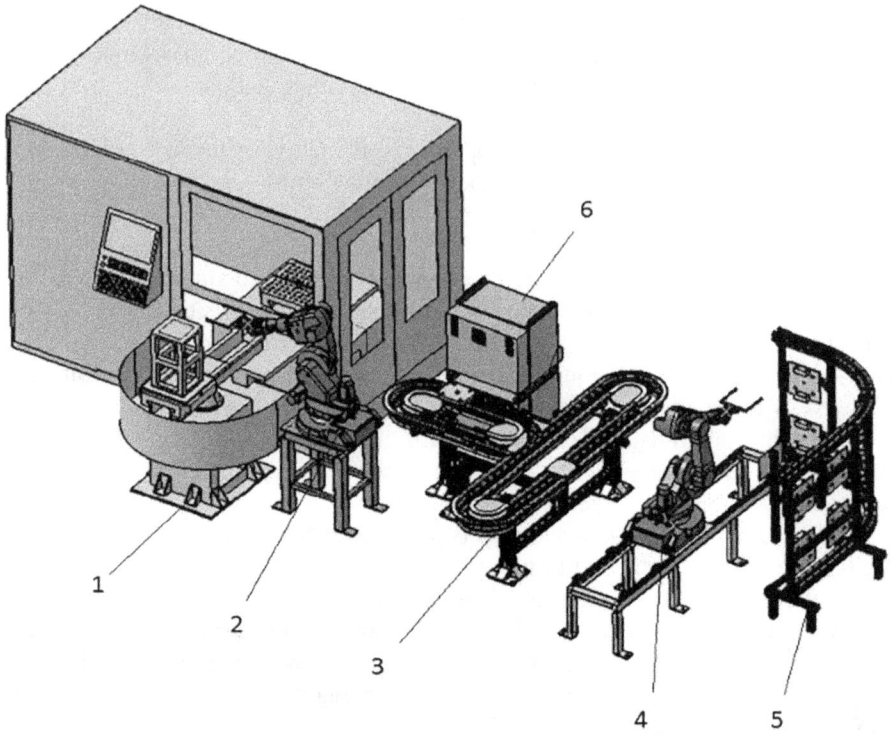

FIGURE 8.1   UO-01-FMC flexible fabrication cell site plan [33].

FIGURE 8.2   UO-01-FMC flexible fabrication cell [33].

In the implementation phase of the digital production were developed several IoT devices could acquire several parameters from some components of the UO-01-FMC flexible manufacturing cell.

The following IoT devices were developed:

- a low-cost one, using a DS18B20 temperature sensor [38]. Possible applications of this device are in development of systems for fire prevention in production halls or other types of buildings. This device triggers an alert in case it detects a temperature that exceeds a pre-set value. The temperatures are acquired at pre-configured time intervals. In the situation where an alert has to be triggered, an HTTPS protocol connects to an API endpoint of a task management software, automatically registering a maintenance task. The task management software automatically informs the maintenance team about the task that was assigned to it. The task management system provides mechanisms to manage the incident with traceability. An experiment in which the newly developed IoT device was validated, the task management system to which it was connected was Pri-Desk [39]. Pri-Desk is a task management system available on the market.

- an IoT device using LM 35 type sensors [30]. It was developed and tested to acquire the bearings temperature values of a three-phase electric motor, having the motor shaft mounted using two radial ball bearings. To monitor the temperature variations of the bearing housings, two LM 35 type sensors were fitted to each bearing.

- an IoT device [40] that measures the voltages and the currents present in the system of the same three-phase electric motor for which the previous IoT device was developed. For this IoT device, the development of three Hall closed-loop current sensors HA2009 and three HV25 ±400 V voltage sensors were used, and as a data acquisition system, a DATA-Q INSTRUMENTS DI-2108 module was used.

- an IoT device [33] that acquires the vibrations of a three-phase electric motor. This IoT device was developed using ADXL345 3-axis accelerometer module and Arduino Nano Compatible Development Board (ATmega328p and CH340). To monitor the vibrations, the accelerometer sensors were placed on the bearings housings.

- an IoT device [41] designed to capture the temperature of the milling tool used by the UO-01-FMC production cell in order to determine its wear. Due to the fact that one goal was to develop a low-cost device, it was used a MLX90640 temperature sensor for data acquisition, in conjunction with a Raspberry Pi 4. The MLX90640 sensor [42] acquires 24×32 temperature values. Using a processing algorithm developed in Python, these values are converted to a thermal image having a resolution of 32×24 pixels. It was decided that the IoT device have this design, due to the fact that it was planned later to analyze the acquired data using specific image classification algorithms. Due to the fact that the used sensor is a low-priced one, it has decided to validate the data acquisition process using a FLIR SC640 thermal imaging camera [43]. The camera acquired images of the milling process at the same time as the MLX90640 sensor. The FLIR SC640 system was not acquiring thermal images but a video of the milling process, a video from which thermal images were extracted.

The next step is to design, develop, and implement a Digital Twin Production application and then to combine the results of the developed digital twin products and production with the usage of the already developed IoT devices that acquire:

- the temperature and vibration of a three-phase induction motor bearing;
- the voltages and currents of the same three-phase induction motor;
- the temperature of a milling tool;

to implement a part of a Digital Twin Performance System related to an automated predictive maintenance system for the UO-01-FMC flexible manufacturing cell.

As part of this last step, were already designed and tested two AI algorithms that can be used with very good results for:

- detection of a faulty operation in the case of a three-phase induction motor [40]. To automatize the decision process if an acquired signal corresponds to a faulty operation, a feedforward artificial neural network was used. For data acquisition, the IoT device using Hall closed-loop current sensors HA2009 and HV25 ±400 V voltage sensors was used. To develop the algorithm, we used MATLAB® and its Machine Learning and Deep Learning modules. The neural network was designed using a two-layer feedforward network. It used a sigmoid transfer function in the hidden layer and in the output layer a Softmax transfer function. A small enough error was obtained after 62 epochs, which is good enough for its purpose.
- for detecting whether the milling tool is worn or not [33]. For data acquisition, the IoT device uses the MLX90640 sensor, which delivers output thermal images with a resolution of 32×24 pixels. Due to this fact, it was decided that the classification algorithm was to be developed based on image classification algorithms. The implemented algorithm was designed using Convolutional Neural Networks. In terms of setting up the training mode, we considered Adam as an optimization algorithm. Five training epochs were considered. After five epochs, the classification accuracy was 100%. We further raised the question: "Would impact the result of classification if images with higher resolution are used?" To find the answer, the acquired images

were replaced with images enhanced to the resolution of 48×64 pixels with the help of Shepard local approximation operators. The response to the question was "YES," since the same classification algorithm was achieved with 100% accuracy after just two epochs.

If we consider the previously presented IoT Devices and classification algorithms, they can be used in several ways to implement a real predictive maintenance system for the UO-01-FMC flexible manufacturing cell and its digital twin production correspondent.

In the real implementation of the predictive maintenance system, the presented IoT devices are used to acquire several parameters which we analyzed using the presented classification algorithms that are detecting if parts of the system, like electric motors, are running properly or faulty, or if the used milling tool is worn or not. In both situations, in case a problem is detected, using an automated integration with Pri-Desk task management system [39], a new maintenance team is recorded and the maintenance team is notified that a new task was assigned to it.

In the implementation of the digital twin production correspondent, the IoT devices are used to feed real data to the digital twin and to analyze if it is performing similarly enough regarding its real correspondent and to fine tune it. Also, the acquired data can be used to develop and train AI algorithms that will later generate virtual data that simulate various types of situations that can occur. Once having access to virtual data that simulates on demand various situations, the already presented algorithms can be further developed, to recognize not only if there is a faulty functioning of an electric motor but also to determine its type, and in the case of the milling tool, to be able to estimate the remaining useful life of the milling tool. Another usage of these virtual data is to design, train, and test new categorization and decision algorithms.

In this way, based on continuous developments of both the predictive maintenance system for the UO-01-FMC flexible manufacturing cell and its digital twin production correspondent, they are fine tuning each other, leading to better and better results.

Further, the Digital Twin Production application development and implementation has to continue, in terms of new IoT devices to be used, in new classification algorithms to be developed, and the design and development of the digital twin of the environment in which the analyzed system is running (Table 8.3).

In Table 8.3, a comparison was made between the steps that have to be performed to implement a FMS and its predictive maintenance system, in both situations: using real equipment versus using digital twins. Some of these steps were already implemented using the results presented in this paper, and the other ones request further research and development. Anyway, an easy conclusion can be formulated: using digital twins, the development is much faster, safer, and cost effective, compared to the conventional approach.

TABLE 8.3

How the Problems Related to the FMS are Solved in Real Manufacturing vs. Using Digital Twin

Type of Problem to be Solved	How the Problem is Solved in Real Manufacturing	How the Problem is Solved Using Digital Twin
Layout Design of Flexible Manufacturing System.	Measurements and tests are conducted to verify the correct posing for each component and FMS as a functional system.	The virtual components are posed in virtual environment and simulations are conducted to check that material flow in FMS could be established.
Developing the program for each FMS component (CNC, robot program)	The CNC and robot' programs are realized online by a human operator, or off-line using CAM software and PLC, robot simulation software, and validated on real FMS components	The CNC and robots' programs are realized using CAM software and PLC, robot simulation software, and validated through simulation software
Testing and validation of manufacturing algorithms in different operating conditions	Different operating conditions are generated/set up using the real workpieces	Different operating conditions are generated/set up using digital workpieces
Testing and validation of safety functions	Dangerous situations can be generated using equipment in the real world and when they occur in the system the safety functions must work properly.	Dangerous situations can be generated using equipment in the digital world and when they occur in the system the safety functions must work properly.
Implementing a predicting maintenance system in FMS	Using IoT devices information related to the FMS components is acquired and using decision software it can be seen if they are running properly or faulty, in the second case using an automated integration task management system a maintenance task is assigned to the maintenance team	The IoT devices are used to feed real data to the digital twin and to analyze if it is performing similarly enough regarding its real correspondent and to fine tune it. The acquired data can be used to develop and train AI algorithms that will later generate virtual data that simulate various types of situations that can occur.

8.4 CONCLUSION

Digital twin technology brings a lot of benefits in developing new products, processes, and services. These benefits are mainly cost reduction, faster development, and flexibility in the development process. Several possible applications of using digital twin technology in the development of a predictive maintenance system were identified. Considering these applications, the solution was to use a digital twin to simulate the environment in which the monitored equipment is running. This way it was obtained in an environment in which the possible errors or interactions can be controlled, thus serving as proper data feeders for the equipment digital twin and for the predictive algorithms.

Encouraging results were obtained by replacing the acquired data with data enhanced with Shepard local approximation operators, as input for the classification algorithms.

In a further development, new machine learning algorithms should be developed to allow the environment digital twin to be trained, and also new algorithms should be considered and tested to validate the input data and to compress/decompress or enhance the transferred data.

REFERENCES

[1] Federal Ministry for Economic Affairs and Energy. Berlin, Germany, "Platform Industry 4.0. (2016). Aspects of the research roadmap in application scenarios.," Jul. 18, 2016. https://www.plattform-i40.de/I40/Redaktion/EN/Downloads/Publikation/aspects-of-the-research-roadmap.html (accessed Oct. 01, 2018).

[2] M. Wollschlaeger, T. Sauter, and J. Jasperneite, "The future of industrial communication: Automation networks in the era of the Internet of Things and Industry 4.0," *EEE Ind. Electron. Mag.*, vol. 11, no. 1, pp. 17–27, Mar. 2017, doi: 10.1109/MIE.2017.2649104.

[3] S. Nižetić, P. Šolić, D. López-de-Ipiña González-de-Artaza, and L. Patrono, "Internet of Things (IoT): Opportunities, issues and challenges towards a smart and sustainable future," *Journal of Cleaner Production*, vol. 274, p. 122877, Nov. 2020, doi: 10.1016/j.jclepro.2020.122877.

[4] R. K. Mobley, *An introduction to predictive maintenance*, 2nd ed. Amsterdam; New York: Butterworth-Heinemann, 2002.

[5] R. Gouriveau, K. Medjaher, and N. Zerhouni, *From prognostics and health systems management to predictive maintenance 1: monitoring and prognostics*. Hoboken, NJ: ISTE Ltd/John Wiley and Sons Inc, 2016.

[6] S. Selcuk, "Predictive maintenance, its implementation and latest trends," *Proceedings of the Institution of Mechanical Engineers, Part B: Journal of Engineering Manufacture*, vol. 231, no. 9, pp. 1670–1679, Jul. 2017, doi: 10.1177/0954405415601640.

[7] A. Grizhnevich, "A comprehensive guide to IoT-based predictive maintenance," *Science Soft*, 2018. https://www.scnsoft.com/blog/iot-predictive-maintenance-guide (accessed Sep. 01, 2021).

[8] "Predictive maintenance with IoT: The road to real returns," *AVNET ABACUS.* https://www.avnet.com/wps/portal/abacus/solutions/markets/industrial/predictive-maintenance-iot/ (accessed Aug. 22, 2021).

[9] M. Pech, J. Vrchota, and J. Bednář, "Predictive maintenance and intelligent sensors in smart factory: Review," *Sensors*, vol. 21, no. 4, p. 1470, Feb. 2021, doi: 10.3390/s21041470.

[10] D. Li, A. Landström, Å. Fast-Berglund, and P. Almström, "Human-centred dissemination of data, information and knowledge in Industry 4.0," *Procedia CIRP*, vol. 84, pp. 380–386, 2019, doi: 10.1016/j.procir.2019.04.261.

[11] C. Franciosi, A. Voisin, S. Miranda, and B. Iung, "Integration of I4.0 technologies with maintenance processes: What are the effects on sustainable manufacturing?,"

IFAC-PapersOnLine, vol. 53, no. 3, pp. 1–6, 2020, doi: 10.1016/j.ifacol. 2020.11.001.

[12] M. Cakir, M. A. Guvenc, and S. Mistikoglu, "The experimental application of popular machine learning algorithms on predictive maintenance and the design of IIoT based condition monitoring system," *Computers & Industrial Engineering*, vol. 151, p. 106948, Jan. 2021, doi: 10.1016/j.cie.2020.106948.

[13] M. Cardona, M. Cifuentes, B. Hernandez, and W. Prado, "A case study on remote instrumentation of vibration and temperature in bearing housings," *JLPEA*, vol. 11, no. 4, p. 44, Nov. 2021, doi: 10.3390/jlpea11040044.

[14] E. Esim and Ş. Yıldırım, "Drilling performance analysis of drill column machine using proposed neural networks," *Neural Comput & Applic*, vol. 28, no. S1, pp. 79–90, Dec. 2017, doi: 10.1007/s00521-016-2322-8.

[15] W. Fontes Godoy, D. Morinigo-Sotelo, O. Duque-Perez, I. Nunes da Silva, A. Goedtel, and R. H. C. Palácios, "Estimation of bearing fault severity in line-connected and inverter-fed three-phase induction motors," *Energies*, vol. 13, no. 13, p. 3481, Jul. 2020, doi: 10.3390/en13133481.

[16] G. Falekas and A. Karlis, "Digital twin in electrical machine control and predictive maintenance: State-of-the-art and future prospects," *Energies*, vol. 14, no. 18, p. 5933, Sep. 2021, doi: 10.3390/en14185933.

[17] "Digital twins and digital twin technology in an industrial context," *I-SCOOP*. https://www.i-scoop.eu/internet-of-things-iot/industrial-internet-things-iiot-saving-costs-innovation/digital-twins/

[18] U. Elangovan, *Industry 5.0: the future of the industrial economy*, First edition. Boca Raton: CRC Press, 2022.

[19] "What is a digital twin?," *Digital Twin Consortium*. https://www.digitaltwinconsortium. org/initiatives/the-definition-of-a-digital-twin/

[20] D. Kinard, "The digital T's -- threads, twins, technology, and transformation," *Bright Talk by TechTarget*, May 04, 2022. https://www.brighttalk.com/webcast/ 18347/542288

[21] "Digital twin technology and simulation: benefits, usage and predictions," *I-SCOOP*. https://www.i-scoop.eu/digital-twin-technology-benefits-usage-predictions/

[22] L. Li, S. Aslam, A. Wileman, and S. Perinpanayagam, "Digital twin in aerospace industry: A gentle introduction," *IEEE Access*, vol. 10, pp. 9543–9562, 2022, doi: 10.1109/ACCESS.2021.3136458.

[23] K. Bruynseels, F. Santoni de Sio, and J. van den Hoven, "Digital twins in health care: Ethical implications of an emerging engineering paradigm," *Front. Genet.*, vol. 9, p. 31, Feb. 2018, doi: 10.3389/fgene.2018.00031.

[24] M. Tyrrell, "Digital twin aids BAE Systems to shape Tempest fighter jet," *AeroSpace Manufacturing*, Aug. 24, 2020. https://www.aero-mag.com/tempest-fighter-jet-bae-systems-24082020/

[25] P. Tucker, "The virtual tools that Built the air force's new fighter prototype," *Defende One*, Sep. 15, 2020. https://www.defenseone.com/technology/2020/09/ virtual-tools-built-air-forces-new-fighter-prototype/168505/

[26] B. Shiklo, "A digital twin approach to predictive maintenance," *Information Week*, Nov. 26, 2018. https://www.informationweek.com/ai-or-machine-learning/a-digital-twin-approach-to-predictive-maintenance

[27] S. Miller, "Predictive maintenance using a digital twin," *MathWorks*. https://www. mathworks.com/company/newsletters/articles/predictive-maintenance-using-a-digital-twin.html

[28] P. Aivaliotis, K. Georgoulias, and G. Chryssolouris, "The use of digital twin for predictive maintenance in manufacturing," *International Journal of Computer*

Integrated Manufacturing, vol. 32, no. 11, pp. 1067–1080, Nov. 2019, doi: 10.1080/
0951192X.2019.1686173.

[29] K. Rahul, "Understand IoT for predictive maintenance in manufacturing," *Software
Advice*, May 28, 2021. https://www.softwareadvice.com/resources/iot-predictive-
maintenance/ (accessed Aug. 25, 2021).

[30] D. Noje, R. C. Tarca, N. Pop, A. O. Moldovan, and O. G. Moldovan, "Automatic
system based on Riesz MV-algebras, for predictive maintenance of bearings of
industrial equipment using temperature sensors," in *Intelligent Methods Systems
and Applications in Computing, Communications and Control*, vol. 1435,
S. Dzitac, D. Dzitac, F. G. Filip, J. Kacprzyk, M.-J. Manolescu, and H. Oros,
Eds. Cham: Springer International Publishing, 2023, pp. 3–19. doi: 10.1007/978-3-
031-16684-6_1.

[31] D. Noje, I. Dzitac, N. Pop, and R. Tarca, "IoT devices signals processing based on
Shepard local approximation operators defined in Riesz MV-algebras," *Informatica*,
pp. 131–142, 2020, doi: 10.15388/20-INFOR395.

[32] D. Noje, R. Tarca, I. Dzitac, and N. Pop, "IoT devices signals processing based on
multi-dimensional Shepard local approximation operators in Riesz MV-algebras,"
INT J COMPUT COMMUN, vol. 14, no. 1, pp. 56–62, Feb. 2019, doi: 10.15837/
ijccc.2019.1.3490.

[33] D. Noje, "Automated system using AI to manage signals transmitted by IoT
devices," PhD Thesis, University of Oradea, 2022.

[34] "Mechatronic Concept Design," *SIEMENS*. https://www.plm.automation.siemens.
com/global/en/products/mechanical-design/mechatronic-concept-design.html

[35] O. G. Moldovan, "Contribuţii aduse la sistemul de gestiune al sculelor în cadrul
celulelor flexibile de fabricaţie. Aplicaţii la celula flexibilă de fabricaţie TMA 55
AL," PhD Thesis, Oradea.

[36] L. S. Csokmai, "Contribuţii privind comanda ierarhizată a sistemelor flexibile de
fabricaţie," PhD Thesis, Oradea, 2013.

[37] Fl. T. Avram, "Contribuţii privind automatizările robotizate la celule flexibile
din industria prelucrătoare a pieselor prismatice," PhD Thesis, Oradea, 2020.

[38] D. Noje, A. Căraban, O. G. Moldovan, O. A. Moldovan, and D. Crăciun,
"Development of DS18B20 temperature sensor IoT device using secure API
connection," *Nonconventional Technologies Review*, vol. 26, no. 3, pp. 39–43,
2022.

[39] "Pri-Desk," *Pri-Desk*. https://pri-desk.ro/ (accessed Jul. 06, 2021).

[40] O. G. Moldovan, R. V. Ghincu, A. O. Moldovan, D. Noje, and R. C. Tarca,
"Fault detection in three-phase induction motor based on data acquisition and ANN
based data processing," *INT J COMPUT COMMUN, Int. J. Comput. Commun.
Control*, vol. 17, no. 3, Apr. 2022, doi: 10.15837/ijccc.2022.3.4788.

[41] D. Noje, O. G. Moldovan, L. S. Csokmai, and A. D. Melinte, "Development of an
IoT device using MLX90640 sensors for temperature acquisition," *Nonconventional
Technologies Review*, vol. Vol 26, no. 4, pp. 44–48, 2022.

[42] "Modul Cameră Termică IR Adafruit MLX90640 24×32," *Optimus Digital*.
https://www.optimusdigital.ro/ro/senzori-senzori-de-temperatura/11185-modul-
camera-termica-ir-adafruit-mlx90640-24x32.html (accessed Aug. 10, 2022).

[43] "Flir SC-320 SC-640 SC-660 high resolution infrared camera for research &
development," *Distek - Measuring Instruments*. http://www.distek.ro/en/Product/
Flir-SC-320-SC-640-SC-660-High-Resolution-Infrared-Camera-for-Research-and-
Development--2061 (accessed Jan. 07, 2022).

9 Artificial Immune-Inspired Disruption Handling in Manufacturing Process

Zubair Ahmad Khan, Ihtisham Ul Haq,
Shahbaz Khan, and Muhammad Tahir Khan

9.1 INTRODUCTION

Today's production systems are significantly more complex and sophisticated than in the past, making them vulnerable to various issues. Process downtime, material loss, efficiency, and production are all negatively impacted by these faults/disruptions. Things like tools malfunctioning, sensors demanding calibration, drive failing, or devices malfunctioning are all examples of such problems. A smart system must be implemented to prevent the aforementioned disruptions, which will directly impact the final product cost [1]. And if not addressed efficiently, they could trigger a chain reaction of subsequent interruptions, causing the entire manufacturing facility to fail catastrophically [2].

The human immune system has inspired the development of comprehensive frameworks for handling disturbances, including the ability to foresee and mitigate any surprises that may arise [3,4]. While there is a variety of literature on fault detection's conceptual framework, specific directions, and procedures for incorporating an Immune-based mechanism into a robotic process are still inadequate. While these methods have been shown to reduce the impact of interruptions, they have seen only limited implementation in actual systems. This research attempts to apply the Fault-Tolerant Framework using an Artificial Immune system (FTFAI), inspired by the human immune system, to a practical manufacturing environment. The suggested methodology could swiftly evaluate interruptions and generate automatic responses. Dynamic evaluation of disruption impacts is achieved by associating weights about fixed states in the database. The chapter is divided into the following sections: 1) Introduction, 2) Human system overview, 3) Literature review, 4) Methodology, 5) Modeling with the HISFDH framework, 6) Discussion, 7) Implementation and results, and 8) Conclusion.

DOI: 10.1201/9781003376620-13

9.2   HUMAN SYSTEM OVERVIEW

The human immune system (HIS) is the body's main defense against harmful external invaders called pathogens. Pathogens, which disrupt the body's normal functioning and lead to illness, are accountable. The HIS is further broken down into the innate and adaptive immune systems, which perform the same functions. The skin, mucous membranes, gastric fluid, epithelial cells, and so on are all examples of barriers that make up the innate immune system. Upon pathogen entry, the body's second line of defense, the phagocytes, engulf the invaders (a process known as "Phagocytosis"), then display a portion of the protein from the pathogen on their surface as a tag, earning them the name "Antigen Presenting Cells" (APC). Antigen Presenting Cells are a part of the innate immune system's ability to recognize and respond to infectious disease (APCs). Antigen Presenting cells (APCs) could tell the difference between "self" and "non-self" thanks to the specific receptors for pathogens known as Pathogens Receptor Cells (PRRs) on the surface of APCs.

The adaptive immune response is the body's time-consuming process of making antibodies against all potential invaders. Both T-cells and B-cells are present in this immune system (Lymphocytes). T-Cells could either serve as a helper cell, stimulating the production and activation of B-cells (cd4), or as a killer cell, destroying any B-cells in their path (cd8). The B-cells number in the tens of billions, and each has a protein called B-cells receptors (BCH) that binds with pathogens via an epitope [1]. After a pathogen binds to a B-cell, the cell undergoes Receptors-mediated endocytosis to digest the invader. Afterward, B-cells attach a particular protein strand (tag) to their outer membrane. APCs are so named because they contain a protein called major histocompatibility complex (MHC) (Antigen Presenting Cells). Interleukin is a signal that APCs produce as a docking site for T lymphocytes. Interleukin triggers B-cell fast mitotic divisions that allow them to deal with several instances of the same infection [2]. The process also involves differentiating some B- and T-cells into B- and T-memory cells, which could recognize the eliminated pathogen. Long-term retention is necessary for effective immune response protection.

The success of the human immune system has sparked interest in developing synthetic immunological defenses (AIS). Antigen identification, functionality involving memorizing, self-arrangement capacity, framing of immune reactions, and pattern recognition capabilities are just some of the AIS-inspired features that researchers have mimicked. The Immune-inspired system also considers reaction-framing features, generalization capability, and multilayer framing [5,6]. Disruption handling, security networks [3,7], milling methods [8], fault detection and diagnostics [9], and network intrusion in computer algorithms [10,11] all use the idea of "Self and Non-Self" to determine the reason of a series of interruptions brought on by malware. Although numerous researchers have taken various approaches to the problem, immunological strategies have also been suggested as a viable solution [12]. [13] provides a thorough discussion of anomaly detection and its practical applications in the industry.

9.3 LITERATURE REVIEW

We've broken up the literature review into three sections: FDI models, alarm management methods, and artificial immune approaches.

9.3.1 MODELS FOR FAULT DETECTION AND IDENTIFICATION IN MANUFACTURING PLANT

The fault detection and isolation (FDI) is commonly employed to deal with disruptions. Two competing models of diagnosis are proposed as the basis for an investigation, with specifics provided for both continuous and logical approaches to handling faults [14]. To cause havoc in the factory, [15] researchers developed a fault tree analysis (a more advanced logical and continuous diagnosis model). In [16], the authors describe an approach for introducing change in the process industry dependent on the hamming distance. A more up-to-date and relevant work [17,18] proposes instantaneously producing a knowledge-based response via PLC configuration and circuit design. Automatic PLC code generation from a failure detection discrete event approach model is proposed in [6]. It also utilizes a special automaton known as non-deterministic output [13] to simulate the control logic. Fault localization is made easier with the help of the techniques presented in [7,8,12,19], which are based on basic manufacturing process designs and use DESs. Modeling control processes employing research based on autonomous automaton [20] Finite state machine (FSM) approach for disruption handling [9] is a major advancement in this area. The sensor and actuator anomaly in the process industry was recommended to be identified by a study centered on exceptional function based on logic [11]. Anomalies in the process industry are revealed by employing the FBMTP technique, as presented in the study by Gosh et al. [11]. The primary goal of this study is to employ a novel and timed response method for fault correction by including the real state transition time into the existing non-deterministic models.

The primary purpose of this study is to employ an innovative and timed response method for fault rectification, taking into account the real state transition time in earlier non-deterministic models. To evaluate on-line degradation and predict the remaining useful life in [21] present a helpful method based on the forward-backward algorithm. Degradation processes with dependent timespan models are proposed. Modeling the processing of sensor information and establishing fault-free operating zones for machines are two examples of the model's processes [22,23]. Process failures could be detected and isolated by employing the model created by Chuang et al. in [17] utilizing modeled nonlinear observers. After identifying the source of the problem, the fault-responsive logic in the system's control system is activated. A study in [24] analyze the costs associated with machine deterioration and propose many preventative maintenance strategies to offset such costs. The correlation between machine failure and quality metrics and the defective rate of products is also examined. Battini et al. found that the mixed integer model positively affected customer value [25]. The manufacturing facility's risks have been reduced due to improvements in assembly line balancing and parts feeding.

TABLE 9.1

Pros and Cons of Agent-Based Manufacturing Systems

Reference	Description	Pros	Cons
[1]	An agent-based production system for the Fourth Industrial Revolution	• Architecture that organizes itself • Dynamically adaptable agents • Dispersed (placed in a remote server farm or the coordinated efforts of agents) • Adaptive scheduling protocol • Organizing in a hierarchy • Using a multi-agent system with the following agents: • Concept, Manufacturing, Transmitting, and Proposal • Undeterred by alterations to the product or external influence, • manufacturing method • produces experimental proof of findings	• This study is limited to the exchange of data and how it relates to the physical world. • No intelligence has been incorporated into the proposed architecture (such as artificial intelligence or bio-inspired techniques)
[2]	Monitoring architecture based on agents for plug and produce-based manufacturing systems	• Topology changes in a network • Utilizes Database • Abstraction (of physical resources), monitoring (to abstract low-level components), and a coordinator are the three agents utilized by a multi-agent-based architecture. • Raw data analysis in a decentralized way • Enhanced performance in monitoring (due to better inter-layer communication protocol)	• An over-reliance on the database, which could be unreliable at times. • There is no way for responding to or correcting any errors that may occur. Only observational efforts are made. • The real-time functionality of the suggested system has not been verified.

In the following Table 9.1, we will examine the two most widely reported agent-based architectures for manufacturing facilities from 2015 through 2022. The approach in this paper provides an alternative mechanism for handling disruptions (including detection and isolation) to determine their origin. A multi-agent architecture that takes its cues from the HIS's inner workings. In contrast to previous methods, a mechanism exists that implements remedial actions according to predetermined criteria. The goal here is to raise the level of productivity in a factory. Both an experimental and simulated environment have validated the suggested architecture.

9.4 LIMITATIONS

In a nutshell, FDI approaches propose a unified nominal control process model to deal with specific interruptions across the production line. Most of these accomplish their aims by utilizing comparison approaches based on log data files of control indicators. Keeping an accurate log of the signals' occurrences is impossible in tightly managed systems with a lot of information flowing through them. But to lengthen the scanning period, such devices require duplicate controllers. This increases the already considerable challenge of providing instantaneous responses in real-time.

9.4.1 ALARM MANAGEMENT

A plant alarm indicates an emergency that needs rapid action. In the instance of alarm, an ontology is established that culdo aid in the investigation of the plant. Da Silva et al. presents an ontology-based relationship of alerts that utilizes regulated data in [18] Advanced monitoring capabilities are a necessary addition to this system, which still has potential for development. In (Quinto and Girardi n.d.), in [26] a filtering technique called SIGARA is employed to construct a model for recommending shares.

On the other hand, it does not provide information on required actions prescribed by the doctor and hence requires context-based modeling. In [27], a general ontology for modeling context is introduced alongside context-aware rules. The idea behind this framework is that it may be applied to various fields.

9.4.2 ARTIFICIAL IMMUNE SYSTEM

The inspiration for artificial immune systems could be traced back to the natural immune system, which serves as a defense mechanism for organisms against diseases (AIS). Antigen (AG) feature identification, pattern memory functionality, self-arranging memory, conversion ability, learning from illustrations, immune reaction framing, spread and simultaneous information processing, multilayer framework, and generalization capacity are all areas where such a system could benefit from further study. AIS could be employed for virtually anything, including but not limited to: optimization, anomaly recognition (including fault diagnosis and computer and industrial network security), robotics, and control (constant and combinatorial). Negative selection, clonal selection, and the immunological network are fundamental to the most important types of AIS. In [25], the authors introduced an AIS with a negative selection mechanism. The idea is to create detectors for change and eliminate any that could single out "Non-self" components. This method has been employed for network intrusion detection [26,28], DNA computers [29], and detecting changes to software caused by viruses [30], milling procedures [31], fault identification and diagnosis [27], and network security [30,31]. Because only cells with the characteristics that AG detects will proliferate, clonal selection is grounded in the principle of essential qualities of adaptive immunity. These copies have deviated from the original cell through mutation at a rate proportional to the adequacy of the match. To solve computational issues, [32,33] developed an algorithm that takes advantage of clonal selection. [34] proposes problems in optimizing both scheduling and the distribution of available resources. CLONALG [35], an algorithm developed by De Castro and Von Zuben, in [9,10] is largely responsible for the spread of the clonal selection technique. You may get your hands on this model's optimization and design identification versions. AINE (Advanced IP. Network Exploitation) developed the artificial immune recognition system (AIRS), another implementation of the clonal selection algorithm. (In the Network Simulator) Immune System Network [36]. Common software for clonal selection is presented in another study [37,38], along with a thorough discussion of the subsequent unimodal, combinatorial, multimodal, and non-fixed function optimization.

There is a discussion of immune concepts and mechanisms in a multi-agent framework that draws on biological immunity notions [3]. To help with reactions to unexpected events, the presented architecture places agents in their environments to process the frequency of interruption and the boundaries of activities. This method involves spotting interruptions, tagging on their direct and indirect effects, and suggesting potential courses of action. However, the AIS-based multi-agent prototype that was produced could not integrate all the necessary parts, nor did it provide a quantitative study of response times.

Further, a study in [39] developed an immune-based framework and methodological recommendations for handling production system disturbances; this made it possible for agent-based technologies to handle such disruptions in a more unified, generic, potent, and responsive fashion. While the proposed method could help resolve the crucial elements However, it is suggested to demonstrate how the existing models may be put into practice and how they could be compatible with cutting-edge tools (like ETA/FTA) and legacy software (such as ERP and MES). To combat these gaps in supply chain and business acumen, researchers [40,41] developed ontologies.

Although the preceding section identifies immune system traits for application in major areas of engineering, the topic of how AIS is employed for disruption handling in the manufacturing sector remains unresolved. Even though academics have developed ontologies for dealing with disruptions in production systems, there is still a great need for further study in this area. Organizations must gain more intelligent and immune properties of AIS through control or automation to better handle interruptions, reduce downtime, and boost productivity.

Models for disruptions are uncommon since nobody seems to care about them. It needs to be given the attention it deserves in the software and domain of production systems. Disruption management insights are rarely formalized since there is little focus on modeling disruption tactics. The offered method addresses gaps in the existing literature by providing an immune-based ontology for the manufacturing process floor shop. A further immune-based disruption handling strategy is conceptualized and executed on a natural system.

9.5 METHODOLOGY

For clarity, this chapter suggests a two-step process.

1. Developing an ontology of the immune systems.
2. Second, progress in immune-based disruption handling.

In the first phase, the process separate into its constituent parts: the actual inputs/tags and the fictitious ones. Then, ontologies are constructed based on identifying digital or tangible tags, and different agents are established. Inputs and outputs, such as drives, valves, actuators, etc., that are hardwired are the true labels. Despite their lack of material form, virtual tags could be essential in facilitating the automation process and assisting the developer. The virtual tags store all the internal alarms, fault bits, reaction bits, analogue sensor ranges, etc.

In the second part of the process, simulate the natural functioning of an immune system to achieve our goals. Human neutrophils and macrophages are the cells responsible for identifying the presence of foreign pathogens in the body. The threat (pathogen) is loaded with enough antibodies to render it harmless after T-cells produce interleukin1, 2, and 3 signals specific to that foreign pathogen and stimulate the exponential clonal proliferation of B-cells. Similarly, the manufacturing facility might be represented using cell agents to ensure that it receives regular data from the control architecture. We use the immune system as an analogy, comparing real data (i.e., Antigen) with simulated data (i.e., APC) and baseline information (i.e., normal body cell). Each scan includes an ongoing comparison of data. The relevant B- and T-agents are informed when an anomaly is discovered. Antibodies produced in this way trigger an appropriate immune response. The database defines this reaction as a list of reactions to a specific occurrence. This procedure adds to the body's already formidable immune system (already in literature). We create a fault-tolerant agent that continuously (with each scan) assesses the abnormality. The fault-tolerant agent intelligently determines the production flow to prevent waste, minimize downtime, and maximize value for the end user based on the relative importance of the problem bits.

9.5.1 ONTOLOGY

Ontologies have emerged to facilitate communication between users, experts, and programmers across all domains [21]. In ontology, the emphasis is on naming and explaining their specific values. This work's immune-based ontology was designed for use on the factory floor, but its applicability extends across the entire production facility.

The created ontology contains manufacturing process factors directly related to the operation of machines/workstations of the floor shop. Still, it does not include information about indirect parameters such as inventory control, worker absenteeism, quality of raw materials, etc. The goal is to include all the control data that's needed to keep the computers running correctly and finish the job.

9.5.2 Cell Agents

Bit matrices stored within cell agents provide comprehensive details about the manufacturing facility. They stand in for the production system's physical and digital components. The data is a core component of any controller/monitoring tool (SCADA or MES) and includes both physical and fictitious components. The manufacturing process is its habitat, and its behavior is information sharing between APC and antigen (Real and Virtual) data. The cell agent's job is to keep the other two agents apprised of its current state so that they could detect and respond to any disturbances.

9.5.3 APC and Antigen Agents

To counteract the antigen agent's disruption matrices, an APC agent provides disruption matrices for every possible virtual disruption. APC and Antigen agent's database contains the sets of potential disruptions, whether actual or virtual, and it also refreshes and stores any unplanned disruption occurring in the manufacturing line. The APC and antigen agent share information with cell agents to keep them up to date. As shown in Figures 9.1 and 9.2, antigen and APC agents serve to identify and characterize any discrepancies between real and digital tags (Figure 9.3).

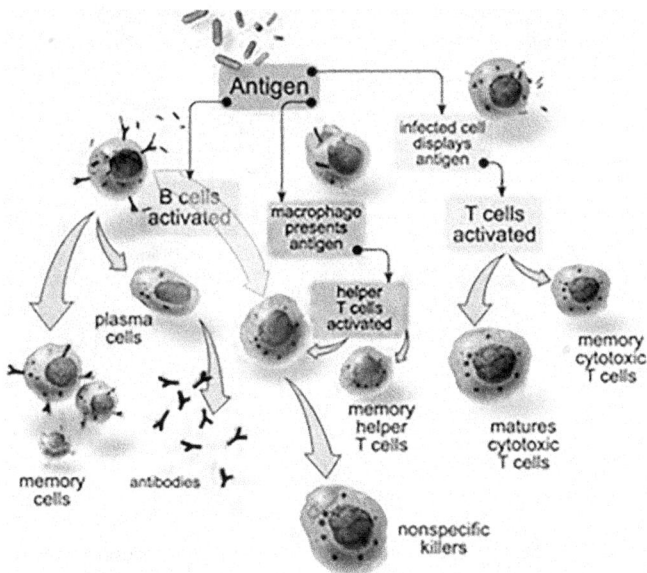

FIGURE 9.1 Overview of the human immune system.

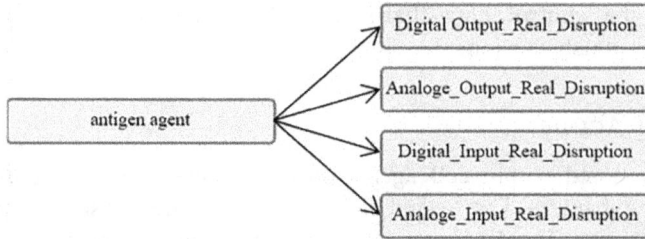

FIGURE 9.2 Protégé antigen agent occurrences.

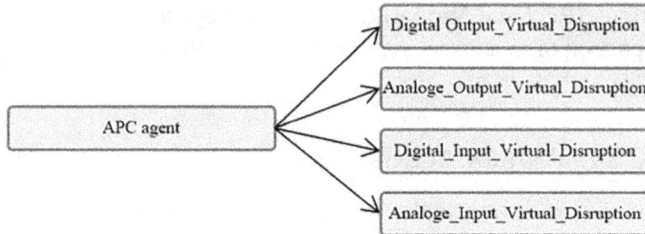

FIGURE 9.3 Protégé APC agent instances.

9.5.4 B- AND T-AGENT

While the B-agent chooses the set of reactions to real-world tag disruptions, the T-agent is responsible for coming up with responses to their virtual-world counterparts. The B- and T-agent database keeps all the corresponding reactions. In the same way, as APC and antigen may, it could refresh its library in the event of new disturbances. The diagrams in Figures 9.4 and 9.5 explain how it works. Once the reactions have been established, they are forwarded to antibody agents.

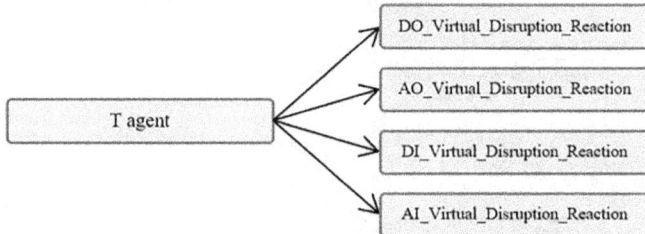

FIGURE 9.4 Protégé's T-agent instances.

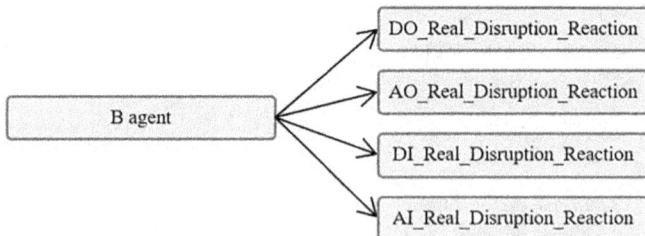

FIGURE 9.5 Protégé's B-agent instances.

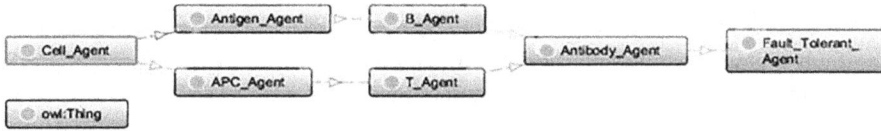

FIGURE 9.6 Evidence of antibody-based agents.

9.5.5 ANTIBODY AGENT

It is the job of the antibody agent to cause the proper response to be made in response to a given disturbance. This model's postulated reaction identification agent.

- The disturbance could be permanent.
- It's possible that the disturbance is mobile and will spread to other stations.

Antibody agents are used to searching through data and filter out unnecessary reactions. Figure 9.5 describes the effects of the B- and T-agents. The diagram shows that the antibody agent is in constant contact with the B- and T-agents to coordinate the most effective response (Figure 9.6).

9.5.6 FAULT-TOLERANT AGENT

By exchanging messages with the antibody agent, the fault-tolerant agent gives relative importance to each possible response based on the current state of events. Cause and effect charts are used to decide the results. The primary goals of the fault-tolerant agent are the minimization of process downtime, the reduction of waste, and the improvement of customer-centric value. Assigning weights to the agent is a customizable feature.

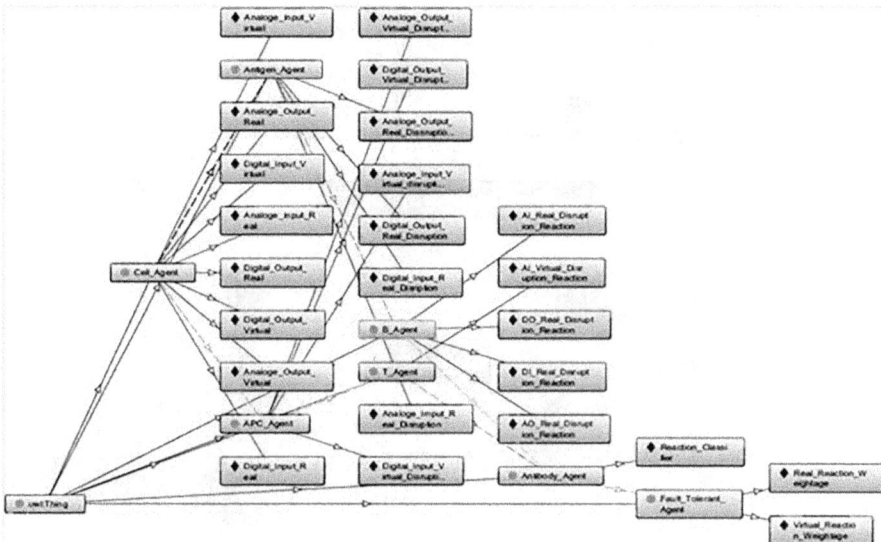

FIGURE 9.7 Agent instances in Protégé.

9.5.7   Fault-Tolerant Agent

The fault-tolerant agent exchanges information with the antibody agent, assigning relative importance to each activity based on the immediate repercussions of taking that action. Cause and effect charts are used to decide the results. The primary goals of the fault-tolerant agent are the minimization of process downtime, the reduction of waste, and the improvement of customer-centric value. To assign different weights in different situations, the agent might be modified.

9.6   MODELING WITH THE HISFDH FRAMEWORK

This method's fundamental idea (shown in Figure 9.8) is to evaluate the predicted behavior of the modeled agents (antigen agent and APC agent) about the actual behavior of the cell agent. Problems are localized as unusual patterns of behavior. Modeled agents (B and T) are compared to sets of known faults, and a response that could tolerate the fault's presence is developed. Antibody agents and weights assigned based on process status are used to process reaction redundancy. The antibody agent processes information, and then the fault-tolerant agent compares the two sets of information to produce the final cell response.

The five-step process outlined later is represented graphically in figure, 1st to Detect, 2nd Identify, 3rd to Evaluate, 4th to Coordinate, and 5th is Validation. A response value is a shape taken by the reaction (RV). This value is then stored in a database and used to determine the most appropriate course of control action to take. There are three potential answers here, and each has an associated RV.

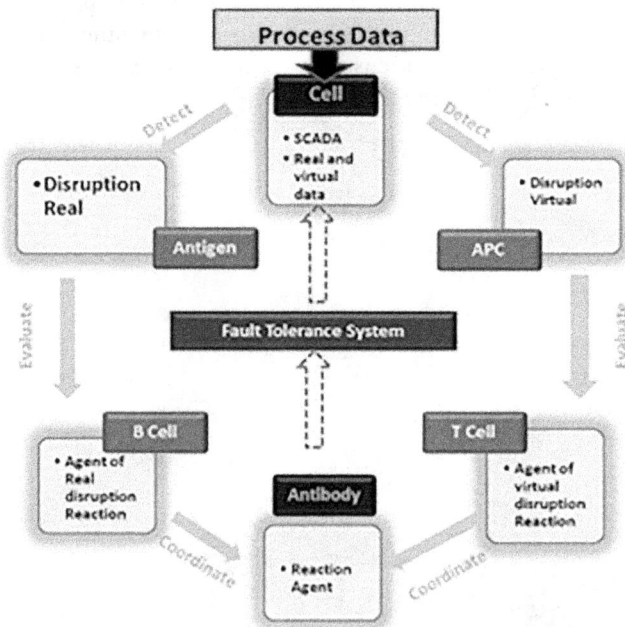

FIGURE 9.8   Proposed methodology.

TABLE 9.2

Tabular Display of Symbols

Symbol	Details	Symbol	Details
T	T agent instances	CAr	Cell agent with real tag
B	B agent instances	Cav	Cell agent with virtual tag
Ap	APC agent instances	ABr	Antibody real
Ab	Antibody instances	Abv	Antibody virtual
Fv	Virtual Disruption	Wab	Antibody weight
Fr	Real Disruption	Ra	Real tag alarm trip point
V	Virtual tag	Va	Virtual tag alarm trip point
R	Real Tag	Wv	Final Weight for virtual disruption
Ws	Workstation	Wr	Final Weight for real disruption

1. Activate the "stopped for critical defects or interruptions" or "stopped" process state.
2. Restart the process and let it run for a while before killing it. This action is effective for medium-intensity faults or disruptions, protecting against sudden process halts to reduce rejection rates.
3. Maintain operation of the procedure. Again, this keeps the production line moving and increases productivity and customer value.

You may find all the information you need about the symbols employed in the suggested procedure in Table 9.2.

9.7 DISCUSSION

9.7.1 DETECTION

The body's immune system uses the key and lock method to identify a foreign infection. In this case, the antigen (a specific protein) on the pathogen's surface allows it to be identified. Each antigen is matched with a unique receptor on the surface of the APCs cells, allowing them to identify and bind to only the relevant pathogens. Here, the information gathered from the factory is applied to this biological phenomenon. Cell agent data describes this information. The cell agent stores real and simulated information, whereas the APC and antigen agents store alarm matrices from the production process (real and virtual). Each tag has a corresponding alert value, as shown in Eqs. (9.4) and (9.5). Data from the cell agent is compared to that from the APC and the antigen agent, revealing the disruption. The control logic is scanned on a per-comparison basis. All data about the process is expressed mathematically in terms of digital and physical labels. To provide for the potential of both sequential and parallel processing, the entire manufacturing facility is broken up into a variety of workstations described by Eq. (9.1)

$$Cell\ Agent,\ CA = \{W_s1,\ W_s2,\ W_s3\ldots\ldots..W_sn\} \tag{9.1}$$

The cell agent further classifies the process tags as either actual or virtual. When n is the total number of workstations, m is the number of virtual and real tags in use, and CAr and CAv are the real and virtual tags, respectively. In addition, the cell agent's columns display the terminals' inputs and outputs (I/Os).

$$CA_r = \begin{bmatrix} r_{11} & r_{21} & . & r_{n1} \\ r_{12} & r_{22} & . & r_{n2} \\ . & . & : & . \\ r_{1m} & r_{2m} & . & rnm \end{bmatrix} \tag{9.2}$$

$$CA_v = \begin{bmatrix} v_{11} & v_{21} & . & v_{1n} \\ v12 & v22 & . & v2n \\ . & . & : & . \\ vm & v2m & . & vnm \end{bmatrix} \tag{9.3}$$

$$Antigen = \begin{bmatrix} ra11 & ra21 & . & ra1n \\ ra12 & ra22 & . & ra2n \\ . & . & : & . \\ ra1m & ra2m & . & ranm \end{bmatrix} \tag{9.4}$$

$$APC = \begin{bmatrix} va11 & va21 & . & va1n \\ va12 & va22 & . & va2n \\ . & . & : & . \\ va1m & va2m & . & vanm \end{bmatrix} \tag{9.5}$$

In this way, the AND gate is used to compare (2) with (4) and (3) with (5), thereby identifying the most likely disruptions that could occur in a production system (see Figure 9.7). Each disturbance could be labeled as either "real" or "virtual."

$$CAr\ XOR\ Antigen\ and\ CAv\ XOR\ APC \rightarrow Fr\ and\ Fv \tag{9.6}$$

The two resulting matrices are real disruption identified (Fr) and virtual disruption detected (Fv). The data is refreshed with every scan cycle and then used in the fault identification process.

9.7.2 IDENTIFICATION

Discovering disruptions is the first step, but selecting appropriate corrective or preventative measures is essential. It is addressed effectively by the proposed model. B-cells secrete antibodies in a biological immune system that kill off infections and damaged cells. T-cells connect to B-cells and communicate with them via interleukin signal, causing the B cell colony to expand. Our method involves the introduction of synthetic B-cells (B-Agents) to detect "antigen agent flaws," while T-cells (T-Agents) are linked to detect APC faults. According to the formulas Eqs. (9.7) and (9.8), the B and T-agents are both one-by-one matrices (8). By just assigning a value of "1" to the faulty tag, they could quickly and accurately pinpoint its location and nature of failure.

$$
B(Cell) = \begin{bmatrix} 1 & 1 & . & 1 \\ 1 & 1 & . & 1 \\ . & . & . & . \\ . & . & . & . \\ 1 & 1 & . & 1 \end{bmatrix} \tag{9.7}
$$

$$
T(Cell) = \begin{bmatrix} 1 & 1 & . & 1 \\ 1 & 1 & . & 1 \\ . & . & . & . \\ . & . & . & . \\ 1 & 1 & . & 1 \end{bmatrix} \tag{9.8}
$$

Dot product of B and Fr, T and Fv with v is the key to fault identification.

$$
Fr \; Dot \; B = Fr . \; B = ABr \rightarrow \text{(Antibody Real)} \tag{9.9}
$$

$$
Fv \; Dot \; T = Fv . \; T = ABv \rightarrow \text{(Antibody Virtual)} \tag{9.10}
$$

In this case, the "actual" antibody is denoted by "ABr," while the "virtual" antibody is denoted by The two matrices created using dot product are given to an antibody agent.

9.7.3 COORDINATION

The antibody's job is to attach to the pathogen, neutralize, or kill it. After determining if a failure is simulated or real, the suggested method moves forward with repair. It is given to an antibody agent, determining whether the problem is static or dynamic. According to Figure 9.7, the discovered disturbances are coordinated between B and T responses. The significance of these disturbances is quantified and awarded a weight. It collects matrices that are the results of the dot product of B and T with the real and virtual fault matrices, as shown in Equation (9.9), and (9.10). Antibody agents could be represented mathematically as a weighted matrix. All manufacturing procedure tags have their related weights. The process's MTTR value is used to create a weight matrix. Mean Time to Repair (MTTR) equals "total downtime" divided by "total repair time" (number of breakdowns). Using MTTR as described, a final weight matrix was (11).

$$
W_{ab} = \begin{bmatrix} W11 & W21 & . & W1n \\ W12 & W22 & . & W2n \\ . & . & . & . \\ . & . & . & . \\ W1m & W2m & . & Wnm \end{bmatrix} \tag{9.11}
$$

As described in Eqs. (9.12) and (9.13), the dot product between the weight matrices (11) and the antibody matrices (9) and (10) are used to derive the weights of only defective tags that occurred during the process (13).

$$ABr \ Dot \ W_{ab} = ABr. \ W_{ab} = W_R \qquad (9.12)$$

$$ABv \ Dot \ W_{ab} = ABv. \ W_{ab} = W_V \qquad (9.13)$$

Antibody weight values (WR and WV) are shown here for both physical and digital tags. These are the final numbers the antibody agent calculated and sent on to the fault-tolerant agent.

9.7.4 FINALIZATION

Fault-tolerant (FT) agent is turned on once the antibody agent reaction is complete. As the first step, the reaction is presented to FT, which then takes in data from the antibody and dynamically evaluates the effects of the reaction. Every action classified as a reaction is given a weight by the FT agent, which it determines based on a threshold it sets. If its value is less than the threshold, it will initiate the response; otherwise, it will choose the next available response. Each workstation's process state, whether active or inactive, is tracked by a fault-tolerant agent. In Eq. (9.14), we could see the Process Status Vector, Ps, as an example of this vector type.

$$Ps \ (Process \ Status) \ = \ \{p1, \ p2, \ ....pn\} \qquad (9.14)$$

Fault-tolerant agents go the extra mile by including the value of malfunctioning computers in their calculations. The workstation's weight vector, Pw, is denoted as

$$Pw = \{W1, \ W2, \ W3 \ldots \ldots ..Wn\} \qquad (9.15)$$

Dot product of Abr with Pw is used to indicate relative importance to affected workstations.

$$Ft \ (Fault \ tolerant) = Abr. \ Pw \qquad (9.16)$$

$$Ft \ (Fault \ tolerant) = Abv. \ Pw \qquad (9.17)$$

Since

$$Abr = \begin{cases} 1, & \text{if faulty workstation} \\ 0, & \text{if workstation has no fault} \end{cases}$$

and

$$\text{Abv} = \begin{cases} 1, & \text{if faulty workstation} \\ 0, & \text{if workstation has no fault} \end{cases}$$

In Eqs. (9.16) and (9.17), the final step involves giving only the malfunctioning computer's weights. The defective tag's weight and the workstation's weight, known as the MTRR value, are necessary components in determining the fault's response value (finalize reaction). The process threshold value for a given workstation's defects is used to determine the response value RV, which is then expressed as

$$RV(i) = (\alpha Ft(i) - \beta W_R(i)) \tag{9.18}$$

The parameter modifies the significance of process weightage and process status. These parameters are fine-tuned for various uses. As the aforementioned process parameters increase, the final tolerance value will be influenced more. Like, a bigger value increases the significance of the tag's influence on the response value. Completeness of the data flow cycle is attained at last. The response is generated as illustrated in Figure 9.8.

As a result, the RVs that emerge from this method fall into one of three broad buckets. The process is immediately shut down whenever a fault's maximum value reaches a certain threshold. This is because, as shown in Eq. (9.18), a faulty station has a much greater impact on the process than a faulty tag, although the latter may be safely disregarded. To put it another way, as soon as this trip point (with a very high critical value) is reached, the system shuts down immediately. When the RV value is at its lowest, it indicates that the defective workstation is of negligible importance and may be safely disregarded. Since the workstation in question has no bearing on the process flow, despite its high importance, it is classified as noncritical. This prompts the observer to respond by continuing the procedure. Some situations may call for a response of "continue the state of operating for some time and then halt it" since the RV value is in the middle of the criticality spectrum. Both a malfunctioning station and a defective tag would have a moderate effect on the entire procedure under these conditions.

9.8 IMPLEMENTATION AND RESULTS

The proposed HISFDH model is applied to a Box Height Sorter Machine as a proof of concept. The Institute of Mechatronics created the modular test bed used for the study. Complete automation and control were implemented with the help of a programmable logical controller (PLC) from Siemens, model S7-1200 (TIA). The control logic in the PLC is developed using portal. As may be seen in Figure 9.9, this apparatus includes a pusher station, a conveyor system, and a sorting device. The primary function of the test bed is to classify the items being

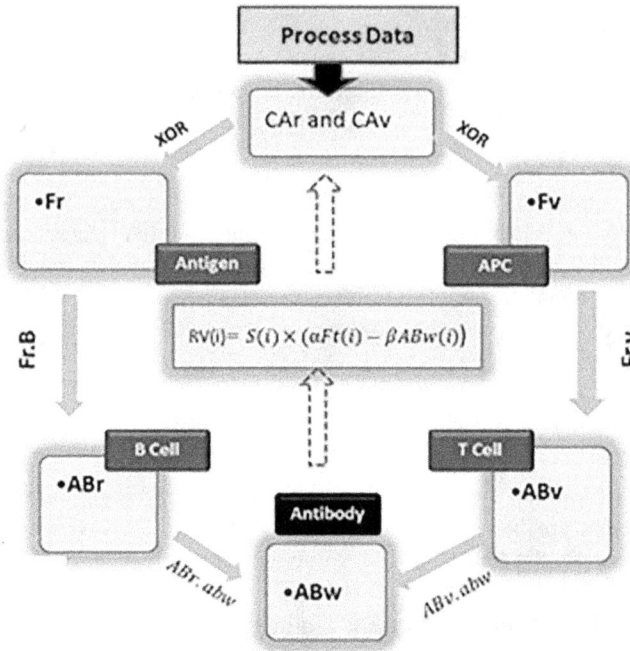

FIGURE 9.9 Immune methods flowchart.

transported by the conveyors. The traditional methods and the HISFDH are incorporated into the process's programming. Protégé is used to create the ontologies. The cellular agent has several inputs and outputs in total. All possible combinations of real and simulated data are loaded into the rig. As was previously mentioned, the antigen and APC agents present templates for actual and virtual tags, and the B and T-agents present reaction bits.

9.8.1 Analyzing Pre-existing Control Logic

The proposed FTFAI model is applied to the Box Height Sorter Machine as a proof of concept. The Institute of Mechatronics developed the modular test bed used in the study. A Siemens S7-1200 PLC was utilized as the controller, and the Integrated Automation (TIA) site was used to program the PLC. This configuration will have a pusher station, a conveyor system, and a sorting device. Sensors could be found at each location in Figure 9.10. The test bed's primary function is to sort the items on the conveyors into predetermined categories. Both conventional methods and the FTFAI are incorporated into the process's code.

Protégé is where the ontologies are created. The cellular agent has several inputs and outputs in total. All possible combinations of real and simulated data are loaded into the rig. As was previously mentioned, the antigen and APC agents present

Section 1 includes pusher motor ,two sensors for position detection with in the box.

Section 2 The Conveyor comprises of the Conveyor Motor (Actuator) and three Proximity sensors (P1- Presence of the product, P2-for sorting the size & P3-for the end seperation). EOCR is used to check the motor current status.

Section 3 Sorter Section comprises of the Sorter Motor (Actuator), and two Proximity Sensors as input.The Proximity Sensors are used to check the current status of the Motor and end separator point to finaly separate the product.

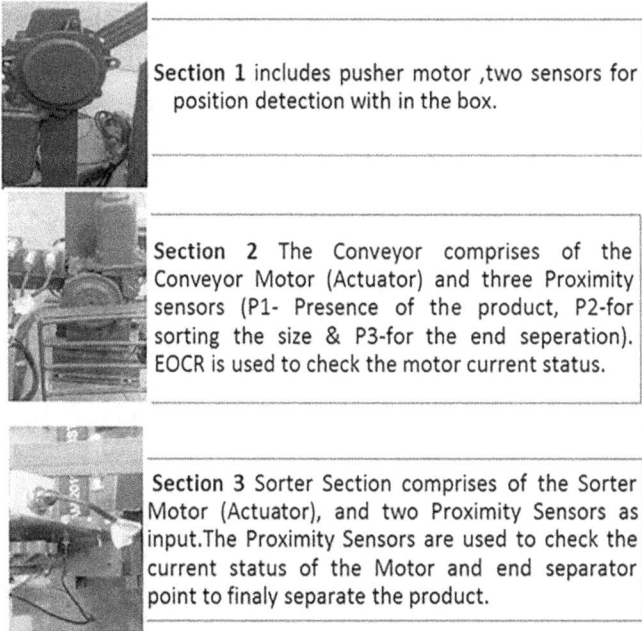

FIGURE 9.10   Testing rig.

templates for actual and virtual tags, and the B- and T-agents present reaction bits. Specifically, Figure 9.10 describes this.

After 24 hours of operation, Figure 9.10 presents a detailed account of the process's downtime. Total process downtime was determined to be 139 minutes (2.12). Overall analysis (shown in the figure) shows that the conveyer end position proximity signal error alarm enable (C.P.P.S.E.A.E) is the primary cause of downtime in workstation 2 (the conveyor section), with the Conveyer Position Proximity Sensor Detect Status (C.P.P.S.D.S) in workstation 1 being the secondary potential parameter, resulting in a 13-minute downtime. Convey Small Proximity Sensor Signal Error Alarm Enable (C.S.P.S.S.E.A.E), with 11 minutes downtime, plays a significant part in disruption. In contrast, Convey Actuator Run Command and Convey Actuator Run status, both of which indicate disruption, are roughly nine minutes. CARS and CAURS suffered outages of nine and seven minutes, respectively.

There are also numerous types of disruption present in the data, which may be found in every tag shown in Figure 9.10. In this analysis, we focus on the top five categories, which account for over 50.0% of interruption downtime. Figure 9.11 and Table 9.2 demonstrate the five most disruptive events that resulted in the longest cumulative downtime (in 24-hour incre-ments). Although many factors contributed to this disruption, the top five are listed above, and they account for half of it. Since they negatively affect

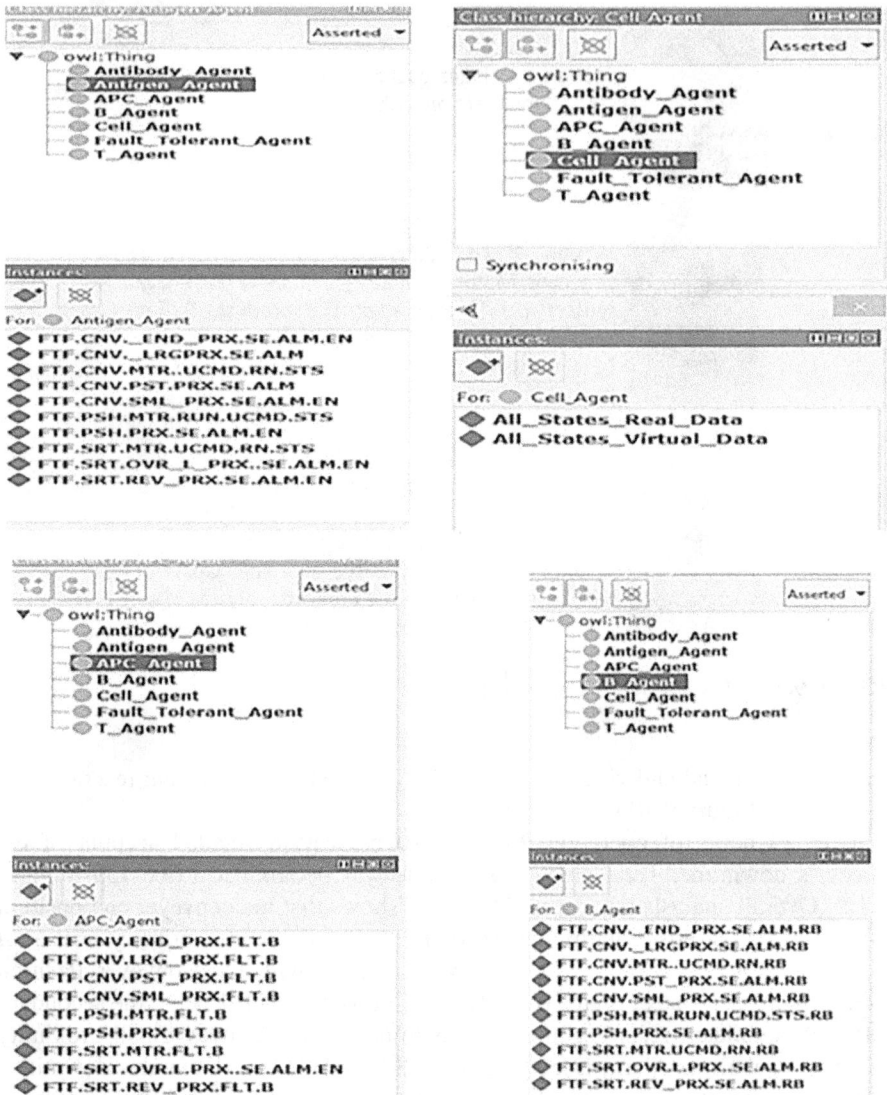

FIGURE 9.11 Instances of agents for use in a test environment.

the proposed model's consumer-centric value and efficiency, they are the model's primary concern (Table 9.3).

9.8.2 MODELING **HISFDH** IN PRACTICE

After integrating the proposed HISFDH model on S7-1200 PLC and WinCC Flexible, we repeat the test bed runs under the same conditions. SCADA software

TABLE 9.3
FBMTP Tag Downtime Analysis

Tags/IOs of three Sections	Symbol	Down Time(minutes)
Conveyer Small Position Proximity sensor Not detect status	C.S.P.P.S.N.D.S	0
Conveyer Small Position Proximity sensor Fault Bit	C.S.P.P.S.F.B	0
Conveyer Small Position Proximity sensor Fault Reaction Bit	C.S.P.P.S.F.R.B	0
Conveyer Small Position Proximity sensor detect status	C.S.P.P.S.D.S	1
Conveyer Position Proximity sensor Not detect status	C.P.P.S.N.D.S	2
Conveyer Actuator Not Run status	C.A.R.N.S	2
Conveyer Small Proximity sensor signal error alarm enable	C.S.P.S.S.E.A.E	2
Conveyer Small Position Proximity sensor detect command	C.S.P.P.S.D.C	2
Conveyer Large Position Proximity sensor Not detect status	C.L.P.P.S.N.D.S	3
Conveyer Large Proximity sensor signal error alarm enable	C.L.P.S.S.E.A.E	3
Conveyer Position Proximity sensor detect command	C.P.P.S.D.C	3
Conveyer Actuator Fault Reaction Bit	C.A.F.R.B	3
Conveyer Position Proximity sensor Fault Bit	C.P.P.S.F.B	3
Conveyer Position Proximity sensor Fault Reaction Bit	C.P.P.S.F.R.B	3
Conveyer End Position Proximity sensor Not detect status	C.P.P.S.N.D.S	4
Conveyer Actuator Fault Bit	C.A.F.B	4
Conveyer Large Position Proximity sensor detect command	C.L.P.P.S.D.C	4
Conveyer Large Position Proximity sensor detect status	C.L.P.P.S.D.S	4
Conveyer Large Position Proximity sensor Fault Bit	C.L.P.P.S.F.B	4
Conveyer End Position Proximity sensor detect status	C.P.P.S.D.S	5
Conveyer End Position Proximity sensor Fault Bit	C.P.P.S.F.B	5
Conveyer End Position Proximity sensor Fault Reaction Bit	C.P.P.S.F.R.B	5
Conveyer End Position Proximity sensor detect command	C.P.P.S.D.C	6
Conveyer Actuator uncommanded Run Status	C.A.U.R.S	7
Conveyer Actuator Run status	C.A.R.S	9
Conveyer Small Proximity sensor signal error alarm enable	C.S.P.S.S.E.A.E	11
Conveyer Actuator Run Command	C.A.R.C	11
Conveyer Position Proximity sensor detect status	C.P.P.S.D.S	13
Conveyer End Position Proximity signal error alarm enable	C.P.P.S.E.A.E	20

with scripting capabilities was used to evaluate the methodology's effectiveness. The proposed framework emphasizes stimulating parameters like C.P.P.S.E.A.E., C.P.P.S.D.S., and C.S.P.S.E.E.E. The test bed was run again for 24 hours. Overall, 82 minutes were recorded (see Table 9.4 and Figure 9.12). This represents a reduction of 50 minutes of downtime. Notable outcomes were seen in the key variables of interest. Similar to how C.P.P.S.D.S. accounted for a loss of 10 minutes (from 13 to 3), C.P.P.S.E.A.E. had the most noticeable reduction of 16 minutes (from 20 to four), while C.S.P.S.E.A.E., C.A.R.C., and C.A.R.S. were noticed with five-minute decreases (from 11 to 6 and 9 to 4). Parameters that had previously contributed less to downtime showed an opposite tendency. For instance,

TABLE 9.4
Implementation Delays in HISFDH Due to Process Downtime

Description	Symbol	Downtime
Conveyer Position Proximity sensor Fault Bit	C.P.P.S.F.B	0
Conveyer Position Proximity sensor detect status	C.P.P.S.D.S	1
Conveyer Position Proximity sensor detect command	C.P.P.S.D.C	3
Conveyer Actuator Fault Reaction Bit	C.A.F.R.B	3
Conveyer Large Position Proximity sensor Not detect status	C.L.P.P.S.N.D.S	3
Conveyer End Position Proximity signal error alarm enable	C.P.P.S.E.A.E	4
Conveyer Actuator Fault Bit	C.A.F.B	4
Conveyer Actuator Run Command	C.A.R.C	4
Conveyer Large Position Proximity sensor detect command	C.L.P.P.S.D.C	4
Conveyer Large Position Proximity sensor detect status	C.L.P.P.S.D.S	4
Conveyer Large Position Proximity sensor Fault Bit	C.L.P.P.S.F.B	4
Conveyer Actuator Run status	C.A.R.S	4
Conveyer End Position Proximity sensor detect command	C.P.P.S.D.C	5
Conveyer End Position Proximity sensor detect status	C.P.P.S.D.S	5
Conveyer End Position Proximity sensor Fault Bit	C.P.P.S.F.B	5
Conveyer End Position Proximity sensor Fault Reaction Bit	C.P.P.S.F.R.B	5
Conveyer End Position Proximity sensor Not detect status	C.P.P.S.N.D.B	5
Conveyer Large Proximity sensor signal error alarm enable	C.L.P.S.S.E.A.E	6
Conveyer Actuator uncommanded Run Status	C.A.U.R.S	6
Conveyer Actuator Not Run status	C.A.R.N.S	7

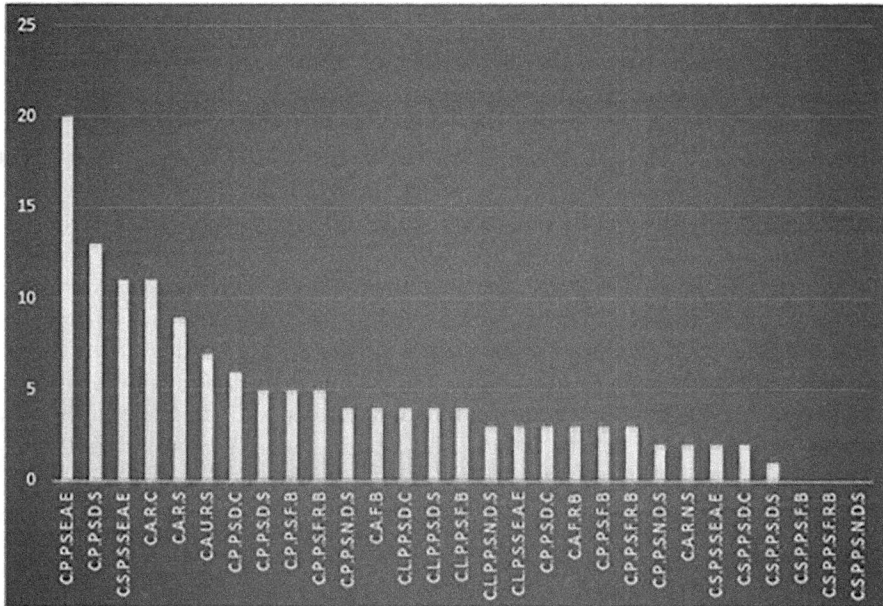

FIGURE 9.12   24 hours process downtime.

C.P.P.S.N.D.S. has increased by four minutes, while C.P.P.S., D.C., and C.L.P.P.S.F.B. have both increased by five minutes. Table 9.4 and Figure 9.12 show that the suggested framework significantly reduces downtime and increases process efficiency and customer-centric value (Figure 9.13).

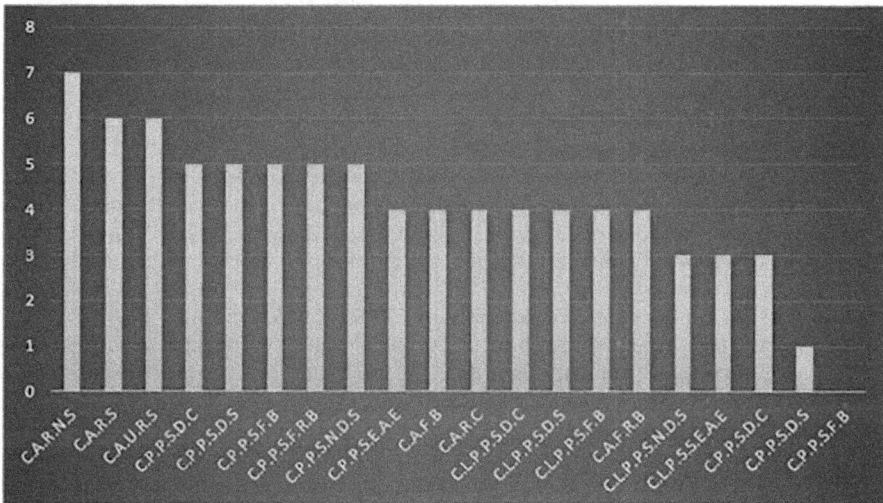

FIGURE 9.13   The HISFDH-induced downtime of a process, in minutes.

9.9 CONCLUSION

This article proposed and demonstrated the use of artificial intelligence systems (AIS) to manage manufacturing disruptions effectively. After ontologies were created, they were put into action (in a lab setting) and analyzed for 24 hours. Positive results for process downtime (24%) were shown in 24 hours for all types of interruptions, and 36 minutes (31%) were seen in five main (targeted) disruptions, according to the results. Further, the proposed HISFDH model led to a 5% boost in process efficiency. A 31% decrease in downtime in the five areas of focus provided an additional productiveness boost. The result was a 5.1% rise in output as a whole. As a result, an additional 216 boxes were sorted by the test apparatus. The potential for the proposed model to improve the plant's efficiency is promising. In the future, the framework could be utilized to enable the unambiguous portrayal of disruption management difficulties for various production factors, including inventory, raw material, and workforce-related issues. The model could be improved further by considering the alignment of workstations along the flow path.

REFERENCES

[1] M. Bruccoleri, Z. J. Pasek, and Y. Koren, "Operation management in reconfigurable manufacturing systems: Reconfiguration for error handling," *International Journal of Production Economics*, vol. 100, pp. 87–100, 2006.

[2] S. Darmoul, H. Pierreval, and S. H. Gabouj, "An immune inspired multi agent system to handle disruptions in manufacturing production systems," in *International Conference on Industrial Engineering and Systems Management, IESM*, 2011.

[3] S. Darmoul, H. Pierreval, and S. Hajri-Gabouj, "Using ontologies to capture and structure knowledge about disruptions in manufacturing systems: an immune driven approach," in *Emerging Technologies & Factory Automation (ETFA), 2011 IEEE 16th Conference on*, 2011, pp. 1–7.

[4] W. Hu, A. Starr, and A. Leung, "Two diagnostic models for PLC controlled flexible manufacturing systems," *International Journal of Machine Tools and Manufacture*, vol. 39, pp. 1979–1991, 1999.

[5] S. Qin and G. Wang, "A study of fault detection and diagnosis for PLC controlled manufacturing system," in *System Simulation and Scientific Computing*, ed: Springer, 2012, pp. 373–382.

[6] A. Machado, D. Lichtnow, A. M. Pernas, A. Bouzeghoub, I. Augustin, L. K. Wives, et al., "A framework reactive and proactive for pervasive homecare environments," in *International Conference on Enterprise Information Systems*, 2014, pp. 320–338.

[7] R. Deaton, M. Garzon, J. Rose, R. Murphy, S. Stevens, and D. Francheschetti, "A DNA based artificial immune system for self-nonself discrimination," in *Systems, man, and cybernetics, 1997. Computational cybernetics and simulation., 1997 IEEE International Conference on*, 1997, pp. 862–866.

[8] D. Dasgupta, "Parallel search for multi-modal function optimization with diversity and learning of immune algorithm," in *Artificial Immune Systems and Their Applications*, ed: Springer, 1999, pp. 210–220.

[9] L. N. De Castro and F. J. Von Zuben, "The clonal selection algorithm with engineering applications," in *Proceedings of GECCO*, 2000, pp. 36–39.

[10] L. N. De Castro and F. J. Von Zuben, "Learning and optimization using the clonal selection principle," *IEEE Transactions on Evolutionary Computation*, vol. 6, pp. 239–251, 2002.

[11] T. Grubic and I.-S. Fan, "Supply chain ontology: Review, analysis and synthesis," *Computers in Industry*, vol. 61, pp. 776–786, 2010.

[12] K. P. Anchor, J. B. Zydallis, G. H. Gunsch, and G. B. Lamont, "Extending the computer defense immune system: Network intrusion detection with a multi-objective evolutionary programming approach," in *First International Conference on Artificial Immune Systems (ICARIS'2002)*, 2002, pp. 12–21.

[13] S. Forrest, A. S. Perelson, L. Allen, and R. Cherukuri, "Self-nonself discrimination in a computer," in *Research in Security and Privacy, 1994. Proceedings., 1994 IEEE Computer Society Symposium on*, 1994, pp. 202–212.

[14] S. Klein, L. Litz, and J.-J. Lesage, "Fault detection of discrete event systems using an identification approach," *IFAC Proceedings Volumes*, vol. 38, pp. 92–97, 2005.

[15] M. Roth, J.-J. Lesage, and L. Litz, "A residual inspired approach for fault localization in DES," *IFAC Proceedings Volumes*, vol. 42, pp. 305–310, 2009.

[16] M. Roth, S. Schneider, J.-J. Lesage, and L. Litz, "Fault detection and isolation in manufacturing systems with an identified discrete event model," *International Journal of Systems Science*, vol. 43, pp. 1826–1841, 2012.

[17] J.-S. Lee and C.-C. Chuang, "Development of a Petri net-based fault diagnostic system for industrial processes," in *Industrial Electronics, 2009. IECON'09. 35th Annual Conference of IEEE*, 2009, pp. 4347–4352.

[18] M. J. da Silva, C. E. Pereira, and M. Götz, "A dynamic approach for industrial alarm systems," in *Computer, Information and Telecommunication Systems (CITS), 2016 International Conference on*, 2016, pp. 1–5.

[19] D. Dasgupta and S. Forrest, "Tool breakage detection in milling operations using a negative-selection algorithm," Technical report CS95-5, Department of computer science, University of New Mexico 1995.

[20] T. Knight and J. Timmis, "AINE: An immunological approach to data mining," in *Proceedings of the 2001 IEEE international conference on data mining*, 2001, pp. 297–304.

[21] A. Azzini, E. Damiani, G. Gianini, and S. Marrara, "An ontology for artificial immune systems," in *Digital Ecosystems and Technologies, 2009. DEST'09. 3rd IEEE International Conference on*, 2009, pp. 324–328.

[22] D. E. O'Leary, "Enterprise ontologies: Review and an activity theory approach," *International Journal of Accounting Information Systems*, vol. 11, pp. 336–352, 2010.

[23] J. Greensmith and S. Cayzer, "An artificial immune system approach to semantic document classification," in *international conference on artificial immune systems*, 2003, pp. 136–146.

[24] E. Hart, P. Ross, and J. Nelson, "Producing robust schedules via an artificial immune system," in *Evolutionary Computation Proceedings, 1998. IEEE World Congress on Computational Intelligence., The 1998 IEEE International Conference on*, 1998, pp. 464–469.

[25] K. Mori, M. Tsukiyama, and T. Fukuda, "Immune algorithm with searching diversity and its application to resource allocation problem," *IEEJ Transactions on Electronics, Information and Systems*, vol. 113, pp. 872–878, 1993.

[26] H. Quintão and R. Girardi, "Especificação dos Requisitos de um Sistema de Gerenciamento de Alarmes baseado na Recomendação de Ações," in *CIbSE*, 2008, pp. 295–308.

[27] S. A. Hofmeyr and S. Forrest, "Architecture for an artificial immune system," *Evolutionary Computation*, vol. 7, pp. 45–68, 1999.

[28] C. J. Matheus, M. M. Kokar, K. Baclawski, J. A. Letkowski, C. Call, M. Hinman, et al., "SAWA: An assistant for higher-level fusion and situation awareness," DTIC Document 2006.

[29] A. Ghosh, S. Qin, J. Lee, and G.-N. Wang, "FBMTP: An Automated Fault and Behavioral Anomaly Detection and Isolation Tool for PLC-Controlled Manufacturing Systems," *IEEE Transactions on Systems, Man, and Cybernetics: Systems*, 2016.

[30] A. Ishiguro, Y. Watanabe, and Y. Uchikawa, "Fault diagnosis of plant systems using immune networks," in *Multisensor Fusion and Integration for Intelligent Systems, 1994. IEEE International Conference on MFI'94.*, 1994, pp. 34–42.

[31] Y. Ishida, "An immune network model and its applications to process diagnosis," *Systems and Computers in Japan*, vol. 24, pp. 38–46, 1993.

[32] J. Bao, H. Wu, and Y. Yan, "A fault diagnosis system-PLC design for system reliability improvement," *The International Journal of Advanced Manufacturing Technology*, vol. 75, pp. 523–534, 2014.

[33] M. Roth, J.-J. Lesage, and L. Litz, "The concept of residuals for fault localization in discrete event systems," *Control Engineering Practice*, vol. 19, pp. 978–988, 2011.

[34] T. Alenljung, M. Skoldstam, B. Lennartson, and K. Akesson, "PLC-based implementation of process observation and fault detection for discrete event systems," in *Automation Science and Engineering, 2007. CASE 2007. IEEE International Conference on*, 2007, pp. 207–212.

[35] H. Aytug, M. A. Lawley, K. McKay, S. Mohan, and R. Uzsoy, "Executing production schedules in the face of uncertainties: A review and some future directions," *European Journal of Operational Research*, vol. 161, pp. 86–110, 2005.

[36] M. Roth, J.-J. Lesage, and L. Litz, "An FDI method for manufacturing systems based on an identified model," *IFAC Proceedings Volumes*, vol. 42, pp. 1406–1411, 2009.

[37] N. Bayar, S. Darmoul, S. Hajri-Gabouj, and H. Pierreval, "Fault detection, diagnosis and recovery using Artificial Immune Systems: A review," *Engineering Applications of Artificial Intelligence*, vol. 46, pp. 43–57, 2015.

[38] W. Hu, M. Schroeder, and A. Starr, "A knowledge-based real-time diagnostic system for PLC controlled manufacturing systems," in *Systems, Man, and Cybernetics, 1999. IEEE SMC'99 Conference Proceedings. 1999 IEEE International Conference on*, 1999, pp. 499–504.

[39] J. Fan, Y. Xie, and M. Ding, "Research on embedded PLC control system fault diagnosis: a novel approach," in *Proceedings of the International Conference on Intelligent Systems Research and Mechatronics Engineering (ISRME'15)*, 2015, pp. 1876–1879.

[40] W. Hu, A. Starr, and A. Leung, "Operational fault diagnosis of manufacturing systems," *Journal of Materials Processing Technology*, vol. 133, pp. 108–117, 2003.

[41] S. Darmoul, H. Pierreval, and S. Hajri–Gabouj, "Handling disruptions in manufacturing systems: An immune perspective," *Engineering Applications of Artificial Intelligence*, vol. 26, pp. 110–121, 2013.

10 Cyber Security in Additive Manufacturing

Muhammad Qasim Zafar, Muhammad Bilal Khan, and Haiyan Zhao

10.1 INTRODUCTION

The automation and IT revolution in industry materialized in the 1970s. The rigorous growth related to network integration, coupling the internet with manufacturing systems, and bringing the idea of the Internet of Things (IoT), has transformed the whole paradigm of manufacturing. Recent advancements in additive manufacturing, particularly at an industrial scale, shifted traditional manufacturing to another paradigm. 3D printing is straightforward for the fabrication of intricate, complex, and customized components directly from a digital file that has changed the entire landscape of manufacturing in a short span of time. Although technology is in its fancy stage perhaps, it has been expediting toward its maturity at an upstanding pace. The efficiency, sustainability, productivity, and quality of additive manufacturing have been greatly elevated through new technologies like Cloud manufacturing [1], Software as a Service (SaaS) [2], Industry 4.0 [3], and real-time service composition [4]. However, the integration of networked devices (cyber systems) and traditional physical manufacturing also enhanced the vulnerability and exposes these systems to new types of risks that were unknown before. Especially additive manufacturing deals with customized, complex, and sophisticated manufacturing which involves a huge digital data transformation from the CAD model to the printing stage that is highly vulnerable to cyber manipulation. The malevolent cyber attacks are of great concern and challenging to detain [5]. In the past decade, several disastrous cases of cyber attacks on manufacturing systems have been reported [6]. Cyber space is the backbone of Industry 4.0, this revolution surges cyber security threats exponentially. Moreover, additive manufacturing (an emerging manufacturing technique) also seems to be compromised due to inadequate cyber physical protection.

10.1.1 Additive Manufacturing and Cyber Security Challenges

Additive Manufacturing (AM) is one of the fastest growing industrial fields in recent years and seems to be limitless with 'any material' (plastics, ceramics, metals, alloy, fibers, high-performance materials, shape-memory materials, etc.), in 'any shape' (solid, hollow, intricate, etc.), applicable to 'any field' (aerospace,

DOI: 10.1201/9781003376620-14

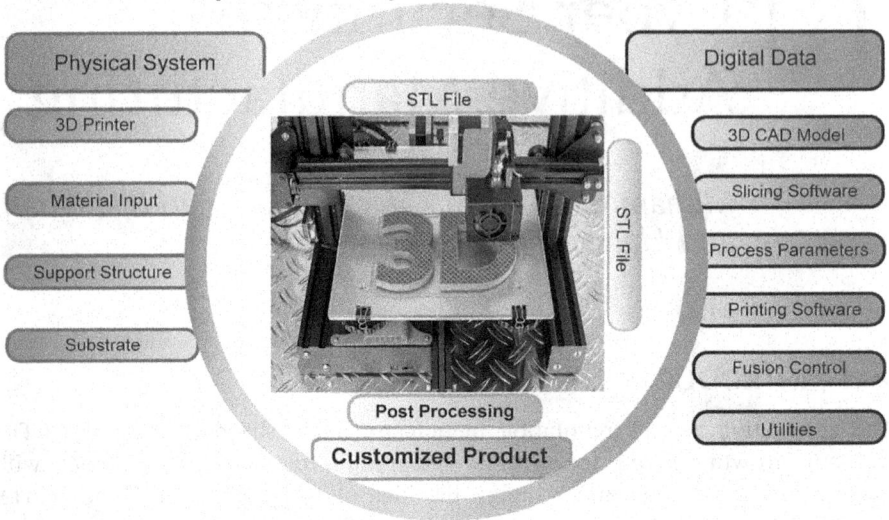

FIGURE 10.1   Cyber vulnerability in additive manufacturing.

biomedicine, automobile, etc.), at 'any workplace' factory, office, home, etc.), and in 'any quantity' (job, batch, mass production) [7]. Initially, AM was considered to be effective for prototyping only because of the time and cost reduction required for molds and dies developing in traditional manufacturing. AM process requires extensive data exchange from a computer to a 3D printer therefore, the biggest challenge so far is to ensure the adequate security of a cyber physical system. The cloud depends on modern production chain concept that is being introduced to achieve cost effectiveness, talent attainability, or work acceleration by splitting it into numerous time zones. This type of distributed cyber physical manufacturing system is highly vulnerable due to various fragments and loopholes. Therefore, companies developing high values industrial components must adopt cyber security precautions to ensure the protection of sensitive data. However, breaches in cyber security are more likely to happen not only for corporates but also for government installations that are expected to be most heavily protected [8]. In such scenarios, it seems that cyber security breaches for the most protected systems, are only a matter of time and the interest of hackers. Such a scenario poses some special considerations for the AM process chain that are vulnerable to cyber-attack. The data flow and directed instructions specified by the software are more susceptible to cyber threats relative to hardware and physical systems in the additive manufacturing cycle (Figure 10.1).

10.1.2   POTENTIAL CYBER THREATS IN ADDITIVE MANUFACTURING

The major AM process components are computers, AM machines (3D Printer), software, and other digital accessories, etc. The connectivity of these components is

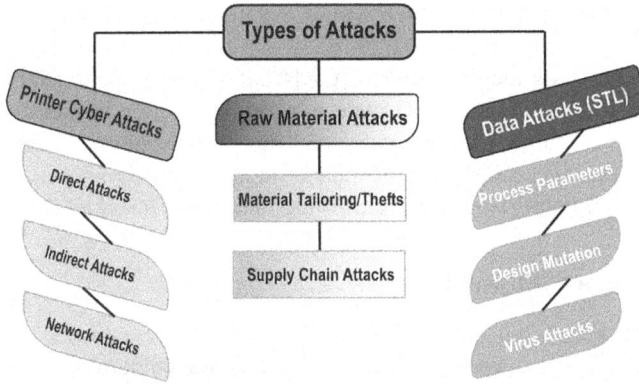

FIGURE 10.2 Classification of cyber security threats in AM.

mandatory for information exchange among components and with the outside world, both physically and wirelessly. Additionally, information exchange could be possible through people or systems. The information flow channels are subjected to different risks. Therefore, the cyber security attack classification in AM is illustrated in Figure 10.2.

In AM, one of the possibilities is to attack the printer. Mainly, side channel attacks are carried out, where attackers use the peripheral system without accessing the printer. As an example, the attacker may hack a smart power meter controlling the power line providing electricity to the manufacturing setup. Hackers can disrupt manufacturing production by tuning or shutting down the power supply. This will eventually disturb the production capacity and can also destroy the printer by altering the power supply continuously. Since 3D printers are meant for remotely monitoring print jobs and managing print queues so this can also be exploited by an attacker. The attacker can intrude through network connectivity and access confidential information related to the printing part easily. To fulfill the purpose of monetary loss for the company. The hacker can purposely induce some defect through a programmed virus in a small number of parts in a large production run. In quality checks, the failure of a few parts leads to the recall of the entire batch, which eventually leads to significant monetary loss. The printer connected to the network can also be attacked directly, where the attack alters the printer parameters like raw material feed rate, extruder temperature, laser power, and other manufacturing conditions to impart a defect in the part. The defects in structural integrity induced by printing parameters are difficult to detect, specifically in large-sized parts [9].

The supply chain attack is mostly planned in two ways. In the first type of attacks, supply chain logistics are being attacked through shipment delays, etc., that eventually affect the industry production schedule. The shipment delay leads to associated consequences such as the expiration of raw material and imminent damage to the packaging such as loss of temperature, moisture barrier, and pressure conditions. In the second type of attack, the intent is to damage the quality of raw materials. The hacker may tune the shipping atmospheric parameters to severely degrade the raw material quality. The processing parameters or raw material manufacturing can be altered, which would eventually degrade the final product of the industry.

A number of steps are involved in CAD solid model, which includes virtual design optimization, slicing, and G-code generation. So, the hacker can attack any of these stages to mutate the design or steal the files for unauthorized production. Similarly, the attacker can use a virus attack to alter the design specifications at any stage. The design data files could be a prime area to be attacked by a hacker.

10.1.3 POTENTIAL MEASURES

We will discuss existing security measures that can be adopted to counter all possible vulnerabilities. To cater to the supply chain vulnerability, a trustworthy environment should be established among diverse businesses in the knowledge of provenance accrued and established for each 3D printed product. The information database should be established to maintain the historical record of each 3D-printed object along with complete information related to adopted additive manufacturing techniques, design, materials, and the supply chain. The centralization of distributed businesses on one platform could be much more effective against tampering attacks. Similarly, the use of blockchain technology for tracking the transfer along with information relevant to the transfer can successfully bring distributed entities to consensus and inhibit tampering attacks. The automatic execution of actions with secure authentication for smart contracts can also be achieved through blockchain technology. Blockchain can provide secure communication among autonomous entities for data sharing through the cloud or networking. The technology architecture can also embed the specific AM industry requirements and its supply chain. It can be altered concerning the transaction capabilities, consensus system, security & privacy, extensibility, and identity management [9].

The possibility of reverse engineering AM products leads to design theft which is challenging to address. To ensure the authenticity and Genuity of the product a serial number, tracking codes, and QR codes should be implanted in AM field. Despite design complexity and variable 3D printed parts, a modern encoding should be applied that can be traceable or read through sophisticated methods like radiography, ultrasound, and/or CT scan. In general, any recognizable signature or method should be devised for identification. Few methods like structure manipulation or slight precipitates orientation or specified targeted areas particularly in a metallic specimen can be tuned via printer settings for identification purposes. The identification of any false or counterfeit part in the market can trigger or caution the manufacturer to look for the possibility of any cyber-attack or design theft. Moreover, machine learning is getting more popular day by day for product authentication and secure virtual environment. The use of AI and machine learning is not yet fully developed but is increasingly being explored in the research community. The evolution of machine learning methods will further enhance AM process control.

Protecting digital files against unauthorized changes or sabotage is challenging. As networking enables a collaborative environment that involves the excessive sharing of CAD files. Therefore, the integrity of transferred data is much more important. The methods have been devised to embed obfuscated design features in the CAD files to ensure security [10]. The security key is required to access the file for printing. The embedment of developed features can be done at any step of AM like in CAD files or STL files, or in the G-code depending on the threat assessment. Similarly, an authentication mechanism should be developed, like a software utility implemented on the printer that can authenticate the specific part for design mutations before printing. The security features are hidden surfaces, surface vector direction in an STL file, slicing angle with respect to x, y, and z directions, and slice thickness. As the CAD files pass through various steps of the process, information loss is expected in the slicing and tool path-creating processes. This particular information that is expected to be lost can be used for developing security features which will eventually work for security enhancement. However, dedicated cyber security professionals should be detained to ensure cloud security and communication channels used for sharing digital data like CAD files, etc. The secure design strategy can provide additional security against the printing of high-quality duplicate parts if the files are stolen by skilled hackers.

10.1.4   CYBER SECURITY AND INDUSTRY 4.0

The industrial sector experienced several revolutions that include mechanization followed by mass production through electricity. Now the current one is the introduction of automation and IT equipment into factories. In 2011, the German government defined the fourth industrial revolution concept, so-called Industry 4.0. The transformation of traditional factories is flexible and more adapted to ever-changing production environments: the Industry 4.0 paradigm, also known as the Industrial Internet of Things or Industrial Internet. Industry 4.0 mainly aims to connect agricultural holdings and manufacturing factories to the Internet.

To establish a smart production environment, disruptive technologies will be needed to ensure communications between all industrial devices throughout the factory and the Internet. These technologies are mainly cloud computing, the Internet of Things (IoT), 3D printing, digital twin, big data, augmented reality, artificial intelligence, and new generation cyber physical systems. The integration of these technologies into the industrial cyber environment makes cyber security considerations mandatory in the design strategy of companies that seek to embrace the Industry 4.0 paradigm [11].

10.1.5   CYBER THREATS IN INDUSTRY 4.0

Cybersecurity characterization is a fundamental aspect of dealing with this. The characterization details are explained in Figure 10.3.

The cyber security threat classification in Industry 4.0 is mainly either technical or managerial. The technical aspect devised the strategy that actually counters the questions: what or who should be protected or from whom the protection is needed and most importantly how it can be protected? The technical team is well equipped with the details regarding cyber criminals and ensures adequate protection methods or techniques to be implemented to counter latest cyber attacks. The technical

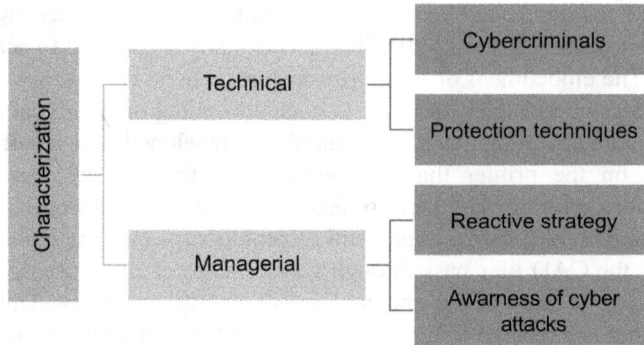

FIGURE 10.3 Cyber threat classification in Industry 4.0.

deficiency that involves lack of technical information or updated protection techniques are usually major loopholes or threats to industry. Therefore, if industry lacks behind technologically, i.e, modern tools or the latest protection techniques are unavailable, the cyber-attack vulnerability increases exponentially, and industry is 100% exposed to attackers. The second main cause of the threat is a lack of proficient managers. The ignorant behavior of industry management toward cyber attacks can lead the industry toward cyber attacks despite the latest protection techniques. The management should be clear regarding cyber security-related issues and must devise a comprehensive and proficient strategy to counter the existing threats. The lack of reactive strategy and awareness plays an integral role in cyber security attacks. These are the two major qualitative areas that usually cause cyber security attacks [12].

10.1.6   Impact of Cyber Security in Industry 4.0

Securing the confidentiality, integrity, and availability of data in a fully connected smart production environment is compulsory where billions of connected IoT devices are communicating and transmitting data. In such an automated and sophisticated environment, if one device is tampered with and manipulated it may affect the entire production life cycle. This can have a catastrophic impact on the whole industrial network. Therefore, cyber security implementation is necessary for Industry 4.0 progress and development.

10.1.7   Cyber Security Countermeasures

Multiple solutions can be opted to confront the cyber security challenges at the industrial level. Generally, two approaches can be implemented; the first one is to take countermeasures to guard against threats or cyber security attacks. This is a short approach to eliminating the existing threat or may limit the harm that an attacker could cause. Countermeasures may enable the mechanism to report or alarm a threat so that corrective action would be taken in response [12]. The fundamental countermeasure at the industrial level is to ensure the safety of the industrial control system. The industrial control system is generally protected using a three-level approach that includes:

- Perimeter isolation i.e implanting firewalls to isolate the plant network from the office network
- Apply multi-layer defense on the network
- Remote user isolation through a dedicated network

This is an active countermeasure to prevent the attack; however, network distribution and area restriction must be implemented within a factory. In modern approaches, encryption is getting popular and can be implemented at desired levels. Communication, stored data, and data stream encryption can be more efficient to counter cyber security attacks. However, adopted countermeasures should be updated periodically so they can be helpful in detecting and preventing cyber threats.

The second one is to dig out and implement the proper solutions for cyber security threats or attacks. One approach is to provide a controlled industrial research environment through protected areas. The ultra-protection should be ensured with limited access to stop the information spread and ensure the homogeneity of the environment. This solution can handle the threats associated with technological upgradation due to its controlled influx nature. Similarly, a possible solution could be to enhance the interconnectivity of exposed devices that could disrupt or infect production and other related systems. The approaches like unidirectional access mechanisms, eradication of access loopholes, and implementation of firewalls can ensure interconnectivity integrity. This solution is also upgradeable with technological advancement. To cater to threats associated with the internet of things (IoT) in Industry 4.0 frameworks like monitoring, classification, and improved security analysis are being focused on. The continuous upgradation and creation of conceptual models are required through software development methodologies. The new functionalities can be incorporated with the mutual consent of the industry and developers to mitigate future challenges. The addition of intelligent response systems in the manufacturing process can also detect and respond to cyber attacks. This can be accomplished by using computational intelligence systems and advanced algorithms like Neural network oracle or artificial neural networks. The machine learning approach can respond to cyber attacks swiftly and smartly. To protect against cyber attacks on low-level industrial control systems like operative sensors or PLC control an intelligent automation knowledge concept is introduced that can specifically PLC, sensors integrated with manufacturing architecture. The ultimate objective of this system is to detect the malicious orders generated by PLC. The detection of malfunctioning at the control level (PLC) eventually leads to the stoppage of manufacturing thus saving the whole system from entering a critical destructive phase.

Knowledge concepts can evolve with evolving technology. Few other solutions eventually immune the industry from cyber threats; that includes honeypots and digital twins. A honeypot is a passive approach devised to collect and analyze

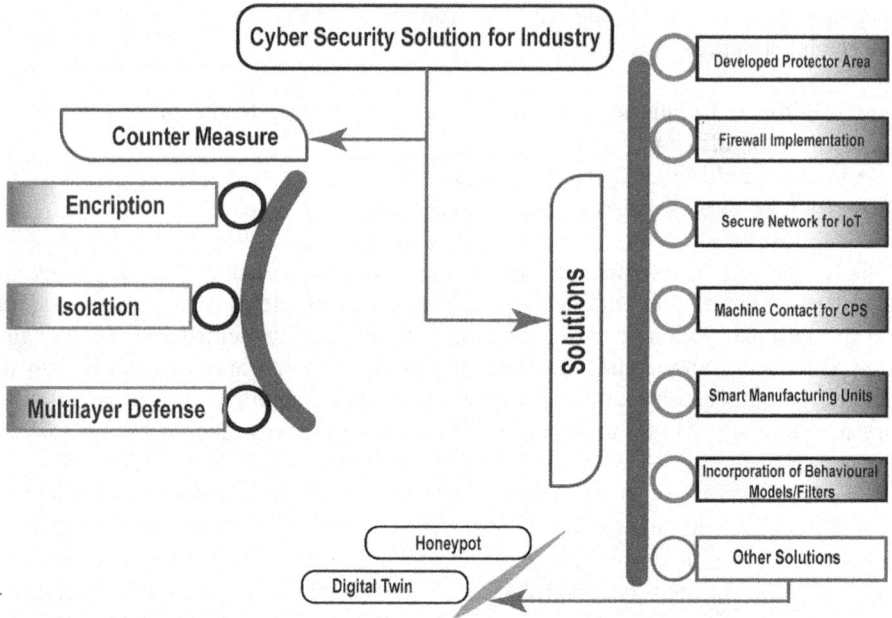

FIGURE 10.4   Cyber security solution for Industry 4.0.

information related to threats or cyber attacks. The core functionality is to alert when the manufacturing vicinity has been breached by hackers. A solution like digital twin is also opted to counter the cyber attacks. The physical objects are replicated virtually to ease the monitoring, visualization, and systems states. The industry is opting and welcoming new approaches and solutions to counter the cyber security threat. The cyber security solutions are summarized in Figure 10.4.

10.2 CASE STUDY

Many cyber attacks have been carried out in the previous few years that caused immense economic loss to well-known companies [13]. To comprehend the cyber-attack feasibility, a simple case study was performed [14]. In a land-grant university, a group of students was asked to perform simple manufacturing tasks to develop tensile test specimens. The reason for choosing a tensile test was its ease of design, machining, and quantifiable performance loss that occurred because of an attack. The main manufacturing steps were to create a CAD model of the specimen, generate the tool path using CAM, transfer the generated tool path to the computer, and finally machine the specimen. The computer was deliberately infected with a virus that altered the specimen dimensions as illustrated in Figure 10.5. There was 19% alteration in the cross-sectional specimen area relative to actual dimensions. The implanted cyber-attack went undetected. If this could be done to any part that would eventually lead to the catastrophic failure-in-use of the product.

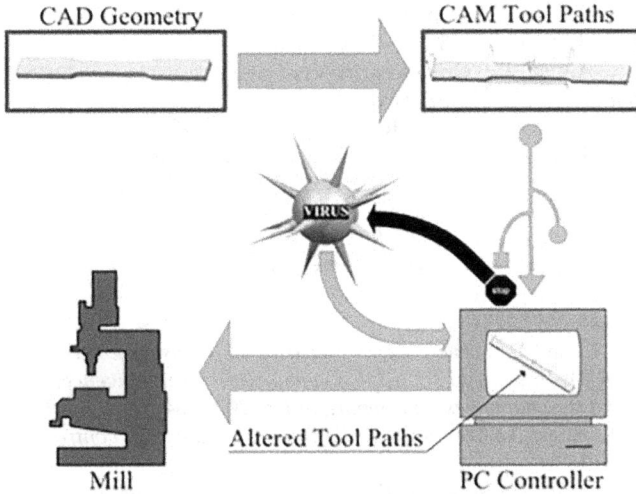

FIGURE 10.5   A case study of cyber security attack.

10.3   CONCLUSION

Cyber security is an integral modern manufacturing aspect, particularly in view of the recent industrial revolution Industry 4.0. Current manufacturing and the smart industrial revolution coupled with recent technologies such as additive and virtual manufacturing are heavily dependent upon digital data, data transfer, data storage, industrial networking, and cloud computing. This data is highly vulnerable to cyber threats, and the biggest future industrial challenge is cyber security, particularly for additive manufacturing. A comprehensive cyber security framework should be developed to counter cyber attacks in modern manufacturing facilities.

ACKNOWLEDGMENTS

This work was funded by the National Natural Science Foundation of China (Grant No. 51975320), the Beijing Natural Science Foundation (No. M22011), and the National Key R&D Program of China (No. 2017YFB1103300).

REFERENCES

[1] Xu, X.J.R. and c.-i. manufacturing, *From cloud computing to cloud manufacturing*. 2012. **28**(1): p. 75–86.
[2] Erisman, R.M., *SaaS adoption factors among SMEs in Indonesian manufacturing industry*. 2013.
[3] Frank, A.G., L.S. Dalenogare, and N.F.J.I.J.o.P.E. Ayala, *Industry 4.0 technologies: Implementation patterns in manufacturing companies*. 2019. **210**: p. 15–26.
[4] Lu, Y. and X.J.J.o.m.s. Xu, *A semantic web-based framework for service composition in a cloud manufacturing environment*. 2017. **42**: p. 69–81.

[5] Turner, H., et al., *Bad parts: Are our manufacturing systems at risk of silent cyberattacks?* 2015. **13**(3): p. 40–47.

[6] Corallo, A., et al., *Cybersecurity challenges for manufacturing systems 4.0: assessment of the business impact level.* 2021.

[7] Wong, K.V. and A.J.I.s.r.n. Hernandez, *A review of additive manufacturing.* 2012.

[8] Padmanabhan, A. and J.J.P.i.A.M. Zhang, *Cybersecurity risks and mitigation strategies in additive manufacturing.* 2018. **3**(1): p. 87–93.

[9] Gupta, N., et al., *Additive manufacturing cyber-physical system: Supply chain cybersecurity and risks.* 2020. **8**: p. 47322–47333.

[10] Gupta, N., et al., *ObfusCADe: Obfuscating additive manufacturing CAD models against counterfeiting.* in *Proceedings of the 54th Annual Design Automation Conference 2017.* 2017.

[11] Thames, L. and D. Schaefer, *Cybersecurity for Industry 4.0.* 2017: Springer.

[12] Bécue, A., I. Praça, and J.J.A.I.R. Gama, *Artificial intelligence, cyber-threats and Industry 4.0: Challenges and opportunities.* 2021. **54**(5): p. 3849–3886.

[13] Setola, R., et al., *An overview of cyber attack to industrial control system.* 2019. **77**: p. 907–912.

[14] Samuel-Smith, L., *A Collective Case Study of Land-Grant Universities and Retention of Students with Learning Disabilities.* 2022: Capella University.

11 Metal Forming in Industry 4.0 Epoch
Challenges and Opportunities

Volkan Esat

11.1 INTRODUCTION

Metal forming constitutes a big portion of the backbone of manufacturing industries across the world, incorporating a great assortment of bulk and sheet processes; cold, warm, and hot work methodologies, and numerous general-purpose and specifically tailored machine tools and presses. At the core of metal forming lies mechanical plastic deformation that is subjected to workpieces in the solid state. The current trends in metal forming include improved mass production capabilities, net and near net shape production, and yielding superior parts in material properties and dimensional requirements. The role of metal forming in the Industry 4.0 (I4.0) epoch is highly worthy of investigation and research, which warrants global efforts to integrate I4.0 to metal forming industries worldwide.

Industry 4.0 brings forward the big idea of smart factories and plants where expertise in various engineering fields such as mechanical, electrical, computer, mechatronics, and software engineering are utilized, leading to a multidisciplinary system thinking that not only opens a road toward development of new technologies but also creates new more effective business models and approaches. The entire effort sparks off innovations that convert conventional factories and manufacturing systems to an entirely connected flexible manufacturing system [1]. Industry 4.0 embodies various critical technologies and approaches in the form of cyber-physical systems (CPS), big data analytics, artificial intelligence (AI), cloud computing and manufacturing, industrial internet of things (IIoT), machine-to-machine communication (M2M), horizontal and vertical integration, autonomous robots, digital twins, simulation, augmented and virtual reality, energy efficiency monitoring, smart maintenance, supply chains, etc. within the smart factory concept that collectively introduce improved autonomy, flexibility, interoperability, predictability, real-time processing, and configurability when compared to conventional factories and practices [1,2].

I4.0 encompasses a variety of techniques, technologies, and approaches. Arguably first and foremost, cyber-physical systems (CPS) comprise intelligent

DOI: 10.1201/9781003376620-15

platforms that are connected through networks including various sensors, actuators, and processors which communicate with the physical entities that complete the functioning of a smart factory. CPS bolster factory operations in real time. Internet of Things (IoT) support cyber-physical systems through Internet via IP addresses, whereas industrial Internet of Things (IIoT) consists of various industrial applications, objects, and devices involving interconnected sensors, instruments, power management systems, and other industrial equipment networked in tandem [1].

Smart factory lies at the core of I4.0 thinking. A smart factory describes an intelligent production environment where manufacturing processes and all related support systems are coordinated mainly without human assistance. Human beings that contribute to proper functioning of I4.0 manifest their contribution in decision making, supervising, and preventive maintenance. In the probable upcoming concept of I5.0, human beings and robots are expected to work together. Industry 5.0 is believed to combine the cognitive abilities that human beings possess such as reasoning, critical thinking, and intuition, along with the strengths and capabilities of robots [1].

The current situation in adaptation of Industry 4.0 shows that the majority of industrial firms and factories globally have not attained a level that I4.0 defines and requires. These firms and factories are observed to have begun installing contemporary digital technologies; however, not fulfilling the entirety of the aims and objectives of the concept of I4.0, yet. In essence, a complete adaptation of Industry 4.0 focuses on establishing a digital infrastructure composed of state-of-the-art digital solutions, which are not yet implemented by many firms and factories. Utilization of new digital technologies might not be as straightforward as it is hoped to be, causing various problems and challenges that can potentially affect the well-being and performance of the companies at every stage. It has been reported that both well-being and system performance are adversely affected during the implementation of contemporary digital solutions, whereas after a successful adaptation, both well-being and performance get better [3].

This chapter aims to address the challenges and opportunities associated with metal forming in the era of Industry 4.0. It is structured first to convey brief information on the past, present, and future of metal forming; followed by Industry 4.0 in the context of metal forming, and the challenges and opportunities associated with it.

11.2 METAL FORMING: PAST, TODAY, AND FUTURE

Metal forming processes as we know it emerge centuries ago through the use of various types of presses in shaping metals. It is one of the earliest families of manufacturing processes in which metals are subjected to plastic permanent deformation for the purpose of attaining desired product shapes and geometries, meanwhile improving material properties, arguably the strength of the material the most. It is estimated that in the fifth millennium BC, a primitive form of forging was applied to shape copper (Cu) into useful form via the application of hammer blows impacting the workpiece. Following advancements included the forging of iron and steel. Although it is predicted that manufacturing processes commenced with the

Stone Age, continued later with Bronze and Iron Ages; breakthroughs in and exponential growth of manufacturing efforts took place after the industrial revolution, reshaping industries and societies [4].

Metal forming manufacturing family is a sizeable one centered around the presses, which are the machine tools that induce permanent deformation to metals for the purpose of generating desired functional products. Presses can be classified based on their operating mechanism in the form of mechanical, hydraulic, and pneumatic presses; and in a different classification regarding their control strategy in the form of conventional and servo types.

Metal formed parts constitute a considerable portion of the manufactured products worldwide today; and it certainly will be a focal area of manufacturing in the years to come, obviously partaking a major role in the Industry 4.0 epoch. Due to its inherent features, various limitations and impediments cause metal forming integration into I4.0 a troublesome period; however, research and development efforts in academia and industry have already begun helping resolve the difficulties associated with it, yielding various new technologies and approaches. Challenges and remedies in metal forming in the era of Industry 4.0 are addressed in the oncoming sections.

11.3 INDUSTRY 4.0 IN THE CONTEXT OF METAL FORMING

It is reported that a central characteristic of the I4.0 revolution happens to be the expanding interaction of parts, machine tools, and human beings. Due to this fact, ever growing mass production capabilities of metal forming needs to be accompanied by factors playing a role in the I4.0 in the form of mass customization and production on-demand. This requirement and need result in an imminent development of new technologies that would transform metal forming into a new mode with greater flexibility. Flexibility in I4.0 metal forming is such a keyword that explains the possible remedies which can be achieved through improved connectivity, enhanced intelligent systems, effective automation, upgraded agility, and scalability. With more flexible metal forming processes, customer needs and requests can be addressed and met promptly [5,6].

The authors claim that the flexibility level can be improved through the following practical methods [5].

- The degrees of freedom can be raised by using flexible, more generic forming tools and by optimizing/changing forming path.
- Employing incremental sheet forming by diversifying the physical process parameters, i.e. temperature, geometry, material type and properties, etc.
- Utilizing various combinations of conventional and flexible forming process, as well as combinations with a non-forming process (i.e. casting)
- Introducing more, enhanced flexibility to the mutual operations of machine, system, and software.

In the successful implementation of metal forming into I4.0, two critical technology notions stand out: integration and connectivity. I4.0 paves the way toward contemporary technological advancements that embody real-time integration, data collection and analysis, transfers among static and dynamic simulations, information transparency, integration of systems possessing specific features, combining real and digital factories, creating awareness, providing technical assistance and interconnection, and remote control of entire facilities along with standalone equipment. Decentralized decision making can be made possible through interoperability of systems and tools granting efficient information management [7,8].

Researchers bring together possible future I4.0 research paths [8]:

- A conceptual framework outlining for a company to realize in order to approach I4.0
- quantitative models and algorithms constituting this conceptual framework while incorporating I4.0 guidelines
- simulations and optimizations of these models and algorithms
- establishing an I4.0 architecture for standardization purposes
- creation of production planning models by consolidating I4.0 enablers
- validation and benchmarking

In the current Industry 4.0 structure, acquisition, analysis, and usage of manufacturing process data plays a considerable role, creating a set of challenges to handle and utilize this big data. The generated data may constitute great sizes, which can be potentially collected during various metal forming processes. Advanced methods need to be formed in order to govern them. As an example, the majority of contemporary metal forming machine tools are servo-controlled such as hydraulic presses for sheet metal forming and swivel bending, and mechanical servo presses for forging operations. Many process variables such as signals and load data that are collected during the processes can be recorded for real-time or further processing and analyses [9].

In another research study applying numerical approaches, the effect of material and friction/lubrication conditions were investigated aiming at estimating the workpiece quality. The authors claim that their approach is a preliminary step targeted toward achieving quality control within an I4.0 framework, which should ideally employ collected data in real time to integrate and enhance the manufacturing process at hand. They identified a prevalent necessity for a paradigm shift from only handling the parameters of forming tools to overseeing features of the end product properties and process conditions [10].

In another research work, authors put forward a framework for improving monitoring and control of sheet metal forming via a hybrid data- and model-based method. The proposed framework employs data from manufacturing stages of the real process to catch and apprehend observations and deviations in order to aid in interpreting this data utilizing knowledge of the process behavior derived from advanced physics simulations. [11]

In a study conducted with an automotive manufacturer, cyber-physical production system (CPPS) capabilities were implemented with minimal infrastructure change [12]. The research focussed on the conditions required for generating a smart Returnable Transit Items (RTIs) system composed of stillages, racks, containers, pallets, and/or cases. The aim is to monitor the flow within the system so that visibility and control of RTI use and management can be enhanced. The designed smart RTI monitoring system upholds management of RTIs usage averting bottlenecks and delays along with quality assurance so that engine defects can be mitigated. The proposed smart RTI framework platform is aimed to help advance future systems with a probable adoption of CPPSs [12].

Sustainability has been a hot topic in all disciplines for a few decades now. The benefits brought by I4.0 such as digitalization of manufacturing processes, more frequent use of smart machines, increasing advanced sensor applications, and incorporation of localized processing units have improved and enhanced many processes in various sectors of industry [4]. From the sustainability perspective; productivity and efficiency have been highly positively affected, wastage has been significantly reduced, digitalized manufacturing processes have been more competent and flexible, and in turn, production has been stimulated and expedited. As discussed earlier, mass production generally causing wastage of resources, and therefore, polluting the environment, may be converted to mass customized production. Emission of harmful polluting gases can potentially be reduced. Improved sustainability of processes and resources as in the case of smart grids with renewable energy resources has demonstrated great potential to attain sustainable smart factories [4]. In other research, authors demonstrated the significant potential of Industry 4.0 for sustainable manufacturing through supporting greater productivity via improving procedures, reducing lead times, and improving corporate efficiency. [13] In an empirical work, authors address companies which are contentiously adopting I4.0 practices at various levels, with expectations to show exponential growth soon, while some of them reaching the conclusion that there is no alternative to higher sustainable performance (SOP) for sustainable growth of a company [6].

It is claimed that I4.0 is the most unsettling concept throughout industrial development, drawing the attention of many countries, sectors, institutions, and organizations toward connectivity, networking, communication, and economic growth [14]. The authors then developed local and global weights in all sub-categories. The organizational (0.413), environmental (0.256), and technological (0.152) obstacles were found to be the top three leading sustainable impediments for digital manufacturing, whereas top management support (ORG1 with GW = 0.2007), leadership (ORG3 with GW = 0.1020), high energy consumption (ENV1 with GW = 0.0696), unstable connectivity among companies (ENV6 with GW = 0.0652), and electronic waste (ENV3 with GW = 0.0607) were spotted as the primary critical sub-categories of sustainable impediments belonging to organizational and environmental impediments groups [14].

In a research focussing on the companies' expectations, it is suggested that organizations should not predict to see significant changes in the way processes are managed and decisions are taken in the short term. They concluded that timing of the technological advances and sociocultural changes implied by I4.0 are not necessarily the same concepts. [15] In another study, researchers' results reveal that various tools such as CPS and IoT and their specific adaptations, such as CPPS and IIoT, prove to be highly beneficial for the introduction of I4.0 to organizations. The aforementioned tools possess the purpose to boost businesses' performance and productivity aiming at more positive results when compared to the competition. Companies choose their set of tools to reach their aims. Authors claim that at the heart of Industry 4.0 lies IoT and Big Data, so called "basic tools" without which a company cannot even contemplate establishing I4.0 practices [16].

11.4   CHALLENGES AND OPPORTUNITIES

In the epoch of Industry 4.0, which industry and academia are both experiencing today, metal forming demonstrates various challenges due to the vast variety of manufacturing systems, materials, products, machine tools, and systems; high cost and low resilience nature of the majority of the processes; probable dynamic changes in friction/lubrication conditions, material properties, and elastic deformations in machines/tools/dies, and also diverse organizational structures and varying company sizes. As stated earlier, focus of industrial production deviated from conventional mass production to mass customization, and with the advancement of I4.0 from mass customization to mass individualization [1,5]. It is apparent that for metal forming processes, augmentation of highly flexible manufacturing procedures, tools, devices, and systems is imperative.

Role of computational modeling and its potential contributions are regarded as essential components of metal forming in the I4.0 epoch. Some researchers claim that finite element analysis (FEA) for sheet metal forming applications continues to be computationally expensive and not adequate to use in real-time closed-loop control. Due to this reason, approximations appear to be a viable alternative for most model-based control efforts for shortening solution timings, whereas designer engineers tend to achieve accurate results rather than faster but poorer in accuracy solutions. Additionally, finite element analysis approach in obtaining approximate solutions is considered to be a demanding method with proper modeling stages, various possible iterations, and validation efforts in order to come up with satisfactory results. Thus, more research and development are required to fully integrate FEA into metal forming within I4.0 [11]. Researchers postulate that integration of digital manufacturing into I4.0 is more beneficial exclusively for those sectors which contain more high-volume manufacturing as in the example of automotive industries in which automation is integrated with networking, ending up with a more flexible manufacturing system [14].

Implementation of optimization arguably seems to be the most universal and auspicious challenge for metal forming within I4.0 by which all collected data can be evaluated and converted into direct instructions for process optimization. Contemporary approaches can be used to solve the problem at hand, and pre-/post-process the generated predictions, after which input parameters can be formed for the metal forming processes. Researchers also stress that effective recycling and novel business models can be facilitated for ecological and economic sustainability [17].

11.5 INDUSTRY 4.0 METAL FORMING TECHNOLOGIES
AND ADVANCEMENTS

In this section, various advancements and technologies for metal forming in the I4.0 epoch are summarized in order to materialize the possibilities for a better adaption of metal forming in the I4.0 context.

- Imparting/improving flexibility in all processes: when it is poor, incremental forming is a remedy, in which the main process is localized deformation. Single point incremental forming is an established dieless technology in this sense [5].
- Incorporation of 3D printing/additive manufacturing utilizing rapid tooling
- Using hybrid approaches such combination of asymmetric incremental sheet forming (AISF) and stretch forming, where a significant downside is material thinning.
- Employing digitized-die forming (DDF), which is a contemporary flexible manufacturing technology that uses specialized machines and tools to transform sheet metal plate into custom irregular 3D shapes. It is also designated as reconfigurable multipoint forming (RMF) which utilizes the concept of a die continuous surface discrete approximation [1].
- Exploitation of flexible media such as elastomers (rubber), liquids (hydroforming), or gases in place of traditional die and punch.
- Utilizing innovative/smart tools which can monitor and control the metal forming processes instead of the traditional passive metal forming processes and tools. With this method, appropriate measures and corrective (active/dynamic compensation) actions can be taken in time so that errors and possible failures can be mitigated or completely prevented from. Through this method, zero defect manufacturing can be achieved [1].
- Applying optical pyrometers, infrared temperature sensors, or thermal imaging cameras in metal forming processes such as forging, in which more accurate and consistent temperature measurement is possible [1].
- Measurement of bend angle and springback through optical (digital imaging, laser optical, light section, etc.) sensors; tactile (rotatable dies, measuring yokes, measuring pins, contact discs, etc.) sensors [18].
- Preferring more flexible, programmable, and simple servo presses: Servo press technology overcomes the many downsides in the respective processes such as materials with low flow ability, excessive heat treatment, resistance in forming, and die and tool design. Servo presses are currently the state-of-the-art to replace conventional metal forming and hydraulic presses with their superior features such as direct driving and toggle drive, fast response rates, and lower power consumption when compared to conventional hydraulic press [4].
- Incorporation of intelligent tools such as contact and non-contact sensors measuring displacement, surface properties, force, temperature,

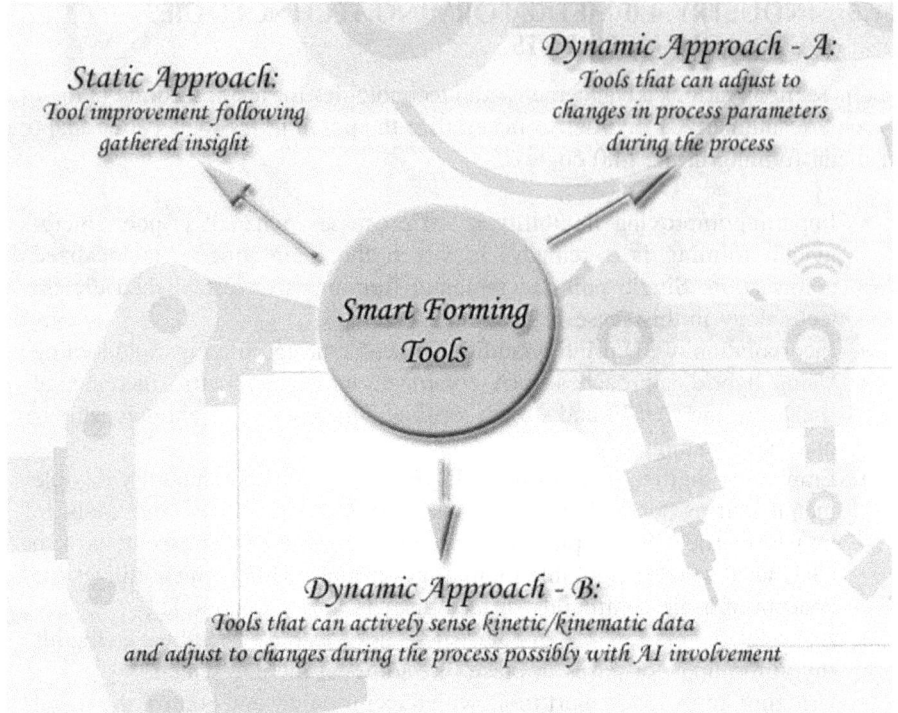

FIGURE 11.1 Smart forming tools.

microstructure, defects, residual stresses, material properties, etc., as well as actuators, and control systems, which possess active, in-process control capability [18]. Utilization of smart forming tools is graphically depicted in Figure 11.1.

11.6 CONCLUSIONS

Metal forming processes in the context of Industry 4.0 are centered around the concept of smart factories embodying smart systems such as flexible forming tools, sensors, and actuators with improved communication, computing, measurement, etc.

This chapter gives some of the most important features and attributes of metal forming in the I4.0 epoch, such as the current state-of-the-art of the approaches, techniques, machine and forming tools, as well as future trends. Flexibility in metal forming is arguably the most decisive objective to be implemented in a smart factory. Flexibility not only involves improved degree of freedom, divergences in physical parameters, combination and hybridization of processes but also makes the operation for machine tool, electromechanical system, and control/analysis software more adjustable.

Finally, sustainability of metal forming processes that are suited to the requirements of I4.0 is discussed, concluding that much more sustainable metal forming solutions are possible through well-implemented I4.0 principles.

REFERENCES

[1] Milutinović, M., Milošević, M., Ilić, J., Movrin, D., Kraišnik, M., Ranđelović, S., and Lukić, D., 'Industry 4.0 and new paradigms in the field of metal forming', Tehnički glasnik, Technical Journal, Vol. 15, No. 2, pp 250–257, 2021.

[2] Karnik, N., Bora, U., Bhadri, K., Kadambi, P., and Dhtrak, P., 'A comprehensive study on current and future trends towards the characteristics and enablers of Industry 4.0', Journal of Industrial Information Integration, Vol. 27, 100294, 11 pp, May 2022.

[3] Kadir, B.A., and Broberg, O., 'Human well-being and system performance in the transition to Industry 4.0', International Journal of Industrial Ergonomics, Vol. 76, 102936, 13 pp, 2020.

[4] Awasthi, A., Saxena, K. K., and Arun, V., 'Sustainable and smart metal forming manufacturing process', Materials Today: Proceedings, Vol. 44, 2069–2079, 2021.

[5] Yang, D. Y., Bambach, M., Cao, J., Duflou, J. R., Groche, P., Kuboki, T., Sterzing, A., Tekkaya, A. E., and Lee, C. W., 'Flexibility in metal forming', CIRP Annals, Vol. 67, No. 2, pp 743–765, 2018.

[6] Gadekar, R., Sarkar, B., and Gadekar, A., 'Investigating the relationship among Industry 4.0 drivers, adoption, risks reduction, and sustainable organizational performance in manufacturing industries: An empirical study', Sustainable Production and Consumption, Vol. 31, pp 670–692, 2022.

[7] da Silva, E. R., Shinohara, A. C., Nielsen, C. P., de Lima, E. P., and Angelis, J., 'Operating digital manufacturing in Industry 4.0: the role of advanced manufacturing technologies, Procedia CIRP, Vol. 93, pp 174–179, 2020.

[8] Canas, H., Mula, J., Diaz-Madronero, M., and Campuzano-Bolarin, F., 'Implementing Industry 4.0 principles', Computers & Industrial Engineering, Vol. 158, 107379, 17 pp, 2021.

[9] Soriani, A., Gemignani, R., and Strano, M., 'A metamodel for the management of large databases: Toward Industry 4.0 in metal forming', Procedia Manufacturing, Vol. 47, pp 889–895, 2020.

[10] Tatipala, S., Pilthammar, J., Sigvant, M., Wall, J., and Johansson, C. M., 'Introductory study of sheet metal forming simulations to evaluate process robustness', In IOP Conference Series: Materials Science and Engineering, Vol. 418, No. 1, 8 pp, IOP Publishing, 2018.

[11] Tatipala, S., Wall, J., Johansson, C. M., and Sigvant, M., 'Data-driven modelling in the era of Industry 4.0: A case study of friction modelling in sheet metal forming simulations', Journal of Physics: Conference Series, Vol. 1063, No. 1, 6 pp. IOP Publishing, 2018.

[12] Neal, A. D., Sharpe, R. G., a van Lopik, K.,a, Tribe, J., Goodall, P., Lugo, H., Segura-Velandia, D., Conway, P., Jackson, L. M., Thomas W. Jackson, T. W., and West, A. A., 'The potential of Industry 4.0 Cyber physical system to improve quality assurance: An automotive case study for wash monitoring of returnable transit items', CIRP Journal of Manufacturing Science and Technology, Vol. 32, pp 461–475, 2021.

[13] Javaid, M., Haleem, A., Singh, R. P., Suman, R., and Gonzalez, E. S., 'Understanding the adoption of Industry 4.0 technologies in improving environmental sustainability', Sustainable Operations and Computers, Vol. 3, pp 203–217, 2022.

[14] Verma, P., Kumar, V., Daim, T., Sharma, N. K., and Mittal, A., 'Identifying and prioritizing impediments of Industry 4.0 to sustainable digital manufacturing: A mixed method approach', Journal of Cleaner Production, Vol. 356, 131639, 20 pp, 2022.

[15] Tortorella, G. L., Saurin, T. A., Hines, P., Antony, J., and Samson, D., 'Myths and facts of Industry 4.0', Int. J. Production Economics, Vol. 255, 108660, 13 pp, 2023.

[16] Gallo, T., Cagnetti, C., Silvestri, C., and Ruggieri, A., 'Industry 4.0 tools in lean production: A systematic literature review', Procedia Computer Science, Vol. 180, pp 394–403, 2021.

[17] Hagenah, H., Schulte, R., Vogel, M., Hermann, J., Scharrer, H., Lechner, M., and Merklein, M., '4.0 in metal forming – questions and challenges', Procedia CIRP, Vol. 79, pp 649–654, 2019.

[18] Allwood, J. M., Duncan, S. R., Cao, J., Groche, P., Hirt, G., Kinsey, B., and Tekkaya, A. E., 'Closed-loop control of product properties in metal forming', CIRP Annals, Vol. 65, No. 2, pp 573–596, 2016.

12 3D Surface Reconstruction of Aluminum Alloy Arc Additive Manufacturing Parts Based on Focused Topography Measurement

Xiaoping Wang and Fengjun Zhang

NOMENCLATURE

CCD	charge-coupled device
SFF	Shape from Focus
GLV	grayscale variance operator
SML	sensitivity of the Summation Laplacian

12.1 INTRODUCTION

Focused evaluation measurement is widely used in industrial, reverse engineering, personalized medical measurement, industrial product quality inspection, cultural relic reconstruction, and restoration. It is mainly an optical system composed of a microscope and CCD. In moving the measured object longitudinally, a CCD camera is used to collect a series of images and record the height information of the images. The evaluation function is used to evaluate the sharpness of pixels and fit the sharpness evaluation value curve to realize the three-dimensional topography measurement of the measured surface.

Shape from Focus (SFF) is an important method for 3D topography measurement. Nayar first proposed it in 1994 [1]. This technique obtains the

DOI: 10.1201/9781003376620-16

depth information of the scene by finding the image position that maximizes the focus evaluation function of the window. Since then, a large number of researchers have continued to study. G. Blahusch *et al.* [2] solved the quadratic surface by least squares fitting to correct the aberration effect of the optical system and improve the topography recovery accuracy. Shim *et al.* [3] proposed a method for filtering the evaluation function value of sequence images based on the zero-phase filter. This method keeps the position of each spatial data point unchanged and eliminates interference to achieve high precision. Pertuz *et al.* [4] proposed a method to compare the performance of different focus measurement operators for focus shapes. Gladines *et al.* [5] introduced a two-step approach to improving shape measurement speed from focus. Mutahira *et al.* [6] used filtering technology to solve the jitter problem of the SFF system and conducted simulation and real experiments to verify it. Ali U [7] presented an alternative method to measure the degree of focus by directly processing colored images. Husna Mutahira [8–11] presentedpresented an alternative method to measure the degree of focus by directly processing colored images. The comparison results showed that the proposed method has the highest correlation and smallest values confirming the effectiveness of using color images for shape recovery. Shape from Focus measurement improves product quality by measuring the surface quality of additive manufacturing. All of the aboveprevious research has focused on proposing focus evaluation algorithms to improve the accuracy and speed of measurement. To more accurately observe the three-dimensional shape of aluminum alloy arc additive manufacturing parts, this study improves the accuracy and stability of the measurement by improving the previous focus evaluation algorithm to obtain more accurate measurement results. The results of 3D topography measurements can provide feedback and improve the quality of the additive manufacturing process.

12.2 SYSTEM AND PRINCIPLE

The CCD camera can collect the reflected light projected by the light source onto the object. The specific light path diagram is shown in Figure 12.1. According to the thin lens imaging principle: when the relationship between the focal length f, the object distance Do, and the image distance D_i is: $1/D_o + 1/D_i = 1/f$, the camera can shoot a clear image, so the plane is called the focal plane. The measurement steps are as follows: (1) Use a CCD camera to collect a series of pictures and record the height information of the images; (2) Use an evaluation function to evaluate the sharpness of the pixel points and fit the curve of the sharpness evaluation value; (3) Realize the three-dimensional topography measurement of the measured surface.

The focus topography recovery technology can be divided into two parts: focus evaluation and scene reconstruction. The reconstruction process is shown in Figure 12.2. An image sequence is a group of acquired sequence images, and the matrix sequence is the image matrix sequence obtained after the focus evaluation of the picture.

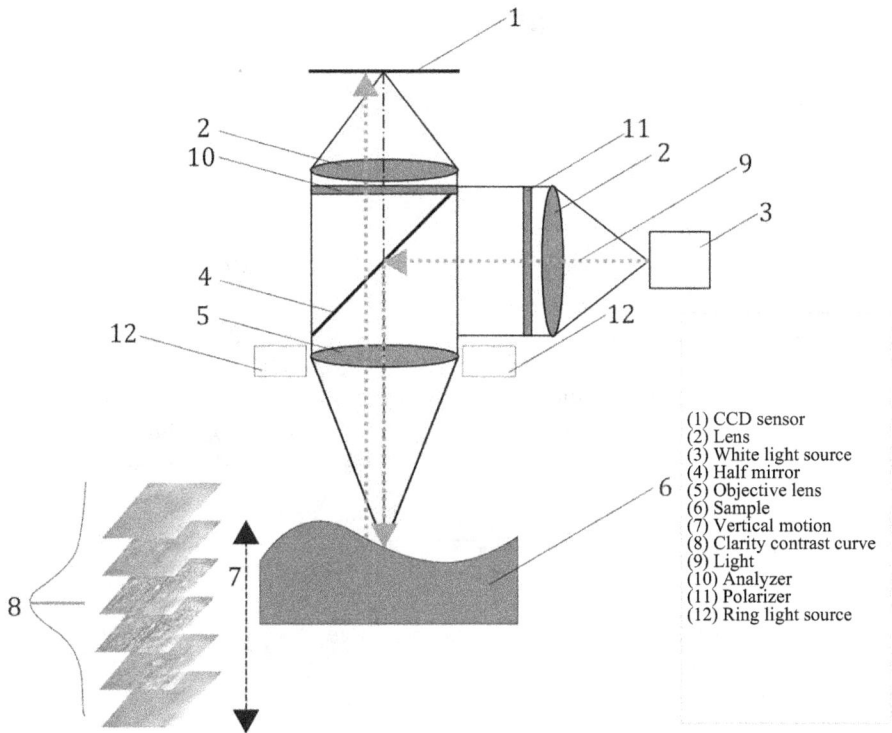

(1) CCD sensor
(2) Lens
(3) White light source
(4) Half mirror
(5) Objective lens
(6) Sample
(7) Vertical motion
(8) Clarity contrast curve
(9) Light
(10) Analyzer
(11) Polarizer
(12) Ring light source

FIGURE 12.1   Schematic diagram of a typical measuring device based on focal length change.

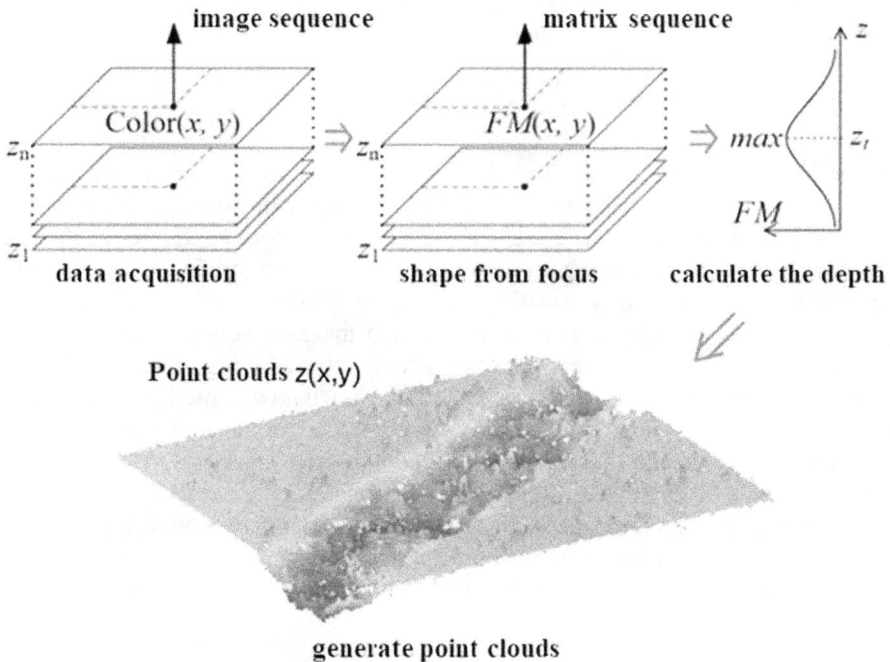

image sequence   **matrix sequence**

$Color(x, y)$ $FM(x, y)$ max

data acquisition   shape from focus   calculate the depth

Point clouds z(x,y)

generate point clouds

FIGURE 12.2   Technical flow of focused topography recovery.

12.3   EXPERIMENTS AND RESULTS

There are many focus evaluation methods to evaluate the clarity of sequence images. Shim *et al.* [3] found that the Laplacian-based focus evaluation operator performed best under normal imaging conditions. Another commonly used focus evaluation operator based on grayscale variance has strong anti-noise performance, but its evaluation accuracy is weaker than that of the Laplacian-based focus evaluation operator. This paper proposes a method combining the statistical-based Gray Scale Variance and the Sum of Modified Laplacian to evaluate the sharpness of the collected images [12]. The low sensitivity of the grayscale variance operator (GLV) to noise and the high sensitivity of the Summation Laplacian (SML) operator to image sharpness are combined to achieve an accurate evaluation of image sharpness. The sharpness evaluation of a point in the GLV is represented by the variance of the gray value of the pixel, and the specific description is as shown in (12.1)

$$FM_{\text{glv}}(x, y) = \frac{1}{n_{(i,j) \in \Omega(x,y)}} \sum_{(I(i,j)-u)^2} \tag{12.1}$$

where n is the total number of pixels in the range $\Omega(x,y)$, and μ is the average value of the pixels in the range.

The following formula can represent SML operator:

$$ML = |2(I(x, y) - I(x - step, y) - I(x + step, y)|$$
$$+ |2(I(x, y) - I(x, y - step) - I(x, y + step)| \tag{12.2}$$

$$F(i, j)_{SML} = \sum_{x=i-N}^{i+N} \sum_{x=j-N}^{j+N} ML(x, y); \quad ML(x, y) \geq T_i \tag{12.3}$$

Where step represents the step size, T_i is the threshold, and N is the field size.

Since the variance is not very sensitive to changes in pixel values, the GLV operator could be more accurate for evaluating sharpness. Its reconstruction accuracy is not high, making its anti-noise ability very strong. This is mainly because the variance is very sensitive to noise. In contrast, the Laplace operator is very sensitive to changes in pixel points, which makes it very sensitive to image definition. In addition, the reconstruction accuracy is very high under low noise, which makes it very sensitive to noise. Usually, a window is used to suppress the influence of noise. Therefore, the study proposes a combined focus evaluation operator by complementing the advantages of the two operators. The algorithm structure is shown in Figure 12.3. These two operators are respectively used to evaluate the sharpness of image sequences and obtain depth data by fitting the focus evaluation value. Compared with the GLV and SML operators, the combined operator reduces the point cloud noise in space while considering accuracy and efficiency.

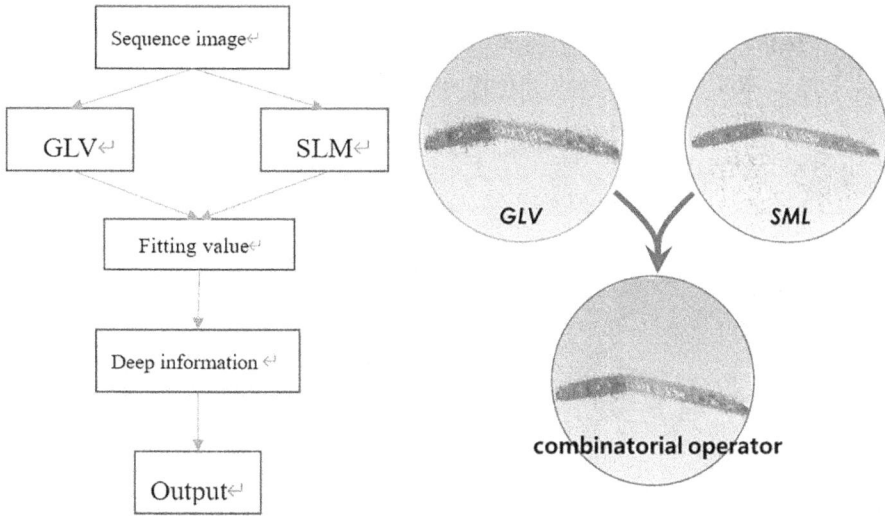

FIGURE 12.3 The algorithm structure.

12.4 ACCURACY AND RELIABILITY VERIFICATION

The accuracy standard block verification experiment platform built in this paper is shown in Figure 12.4. Aliconna's standard step block was used for testing, and the experimental results are shown in Table 12.1. The measurement accuracy of the system under 10' magnification is about 742.9 nm, while the error under 5' magnification is significantly larger than that under 10' magnification, which is 6.44 μm. The larger the objective lens magnification, the higher the accuracy.

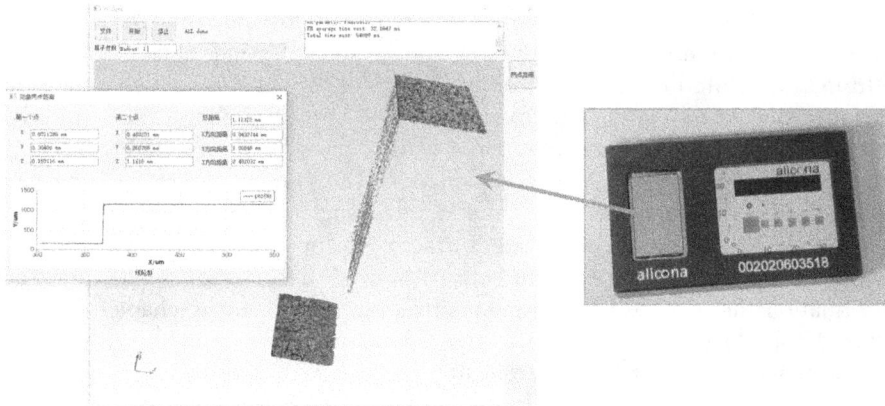

FIGURE 12.4 The accuracy standard block verification experiment platform.

TABLE 12.1

Comparison of Measurement Results with Standard Accuracy Block

Group	μm	μm	μm
10X_1	999.4587	1000.0000	−0.5413
10X_2	1000.9064	1000.0000	0.9064
10X_3	1000.8184	1000.0000	0.8184
10X_4	1000.2564	1000.0000	0.2564
10X_5	1000.3951	1000.0000	0.3951
5X_1	1004.0015	1000.0000	4.0015
5X_2	1007.5386	1000.0000	7.5386
5X_3	1007.5386	1000.0000	7.5386
5X_4	1004.818	1000.0000	4.8180
5X_5	1007.9154	1000.0000	7.9154

12.5 3D MORPHOLOGY RECONSTRUCTION OF ALUMINUM ALLOY ARC ADDITIVE MANUFACTURING PARTS

3D printing technology is widely used due to its unique modeling advantages and has a wide range of applications and development prospects in industrial and personalized medical customization. However, surface defects such as surface roughness, over-melting, and shrinkage cavities are still the factors restricting the development of metal 3D printing [13]. High-precision measurement of surface topography can not only evaluate the surface quality of metal 3D printed products but also provide experimental data for improving the manufacturing process. On the other hand, the surface conditions of metal 3D printed products are highly complex, and many surface topography measurement methods are difficult to measure surfaces with too much roughness. Focus microscopy can measure particularly rough surfaces and obtain true color information while measuring the object's surface topography, which can detect surface shrinkage, unmelted oxidation, and other defects. Arc additive manufacturing of aluminum alloy thin-walled parts is shown in Figure 12.5. The processing process is accompanied by high-temperature melting and solidification of the metal wire, and defects such as shrinkage cavities and porosity are prone to occur on the surface. 3D topography reconstruction by acquiring sequence images. The clear part of the image is the practical information that can be used to reconstruct the image in this chapter, and the blurred part does not participate in the reconstruction.

The sequence images in the acquisition process are shown in Figure 12.6. The sequence images in the figure are the 1st frame, the 39th frame, the 81st frame, and the 181st frame, respectively.

The 3D topography reconstruction result can be obtained by processing the collected data, as shown in Figure 12.7. Note that the color information carried by

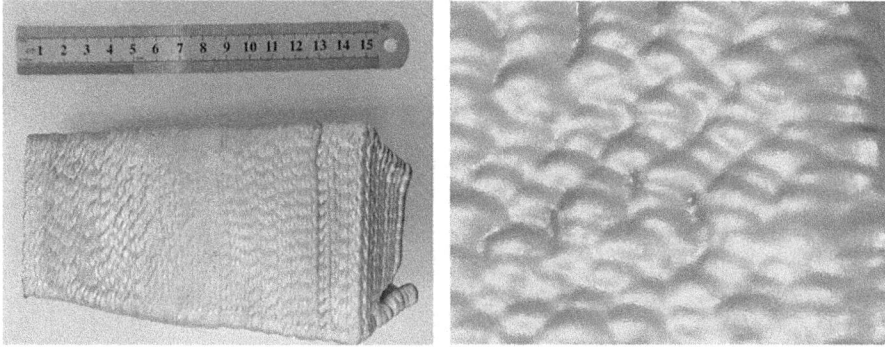

FIGURE 12.5 Surface topography of aluminum alloy thin-walled parts fabricated by the arc and 3D structured light measurement.

FIGURE 12.6 Sequence images acquired during the reconstruction.

the ordered point cloud is the real color information. It can be seen from the displayed three-dimensional morphology that the surface of aluminum alloy arc additive manufacturing parts is prone to shrinkage cavity, which is due to the defects formed by uneven cooling at high temperatures.

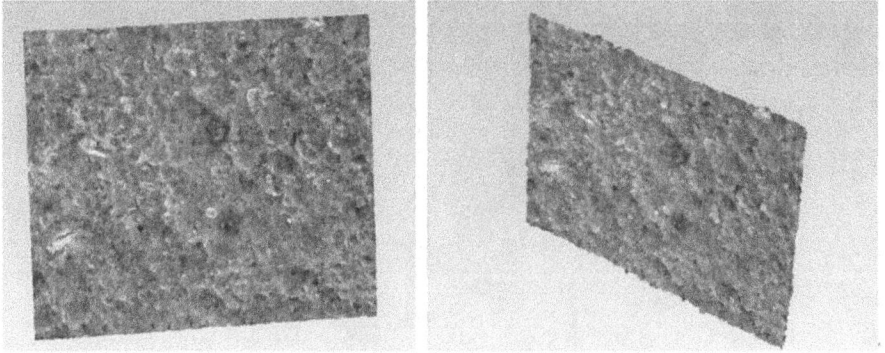

FIGURE 12.7 Front view and isometric view of the surface reconstruction of 3D topography of aluminum alloy arc additively manufactured parts.

12.6 CONCLUSION

This paper proposes a method combining the statistical-based gray variance operator and the Laplacian sum operator to evaluate the sharpness of the collected images. The low sensitivity of the GLV to noise and the high sensitivity of the SML to the image sharpness are combined to achieve an accurate evaluation of image sharpness. Experiment results show that the proposed method can effectively fill missing points to obtain complete point cloud data. According to the principle of focused topography recovery technology, a complete set of measurement hardware equipment is built, and the surface topography measurement of the aluminum alloy arc additive manufacturing parts is carried out. Finally, a reliable 3D topography measurement method has been developed.

REFERENCES

[1] Nayar S K. Shape from focus system[C]//PrGenerate a point cloudoceedings 1992 IEEE Computer Society Conference on Computer Vision and Pattern Recognition. IEEE, 1992: 302–308.
[2] Helmli F S, Scherer S. Adaptive shape from focus with an error estimation in light microscopy [C]// Proceedings of the 2nd International Symposium on Image and Signal Processing and Analysis, Pula, Croatia, June 21, 2001: 188–193.
[3] Shim S O, Malik A S, Choi T S. Noise reduction using mean shift algorithm for estimating 3D shape[J]. The Imaging Science Journal, 2011, 59(5): 267–273.
[4] Pertuz S, Puig D, Garcia M A. Analysis of focus measure operators for shape-from-focus[J]. Pattern Recognition, 2013, 46(5): 1415–1432.
[5] Gladines J, Sels S, Blom J, et al. A fast shape-from-focus-based surface topography measurement method[J]. Sensors, 2021, 21(8): 2574.
[6] Mutahira H, Shin V, Muhammad M S, et al. Sampling-noise modeling & removal in shape from focus systems through Kalman filter[J]. IEEE Access, 2021, 9: 102520–102541.
[7] Ali U, Lee I H, Mahmood M T. Guided image filtering in shape-from-focus: A comparative analysis[J]. Pattern Recognition, 2021, 111: 107670.

[8] Mutahira H, Ahmad B, Muhammad M S, et al. Focus measurement in color space for shape from focus systems[J]. IEEE Access, 2021, 9: 103291–103310.

[9] Smid P, See C W, Somekh M G, et al. Non-iterative aberration retrieval based on the spot shape around focus[J]. Optics and Lasers in Engineering, 2022, 157: 107108.

[10] Ali U, Mahmood M T. Depth enhancement through correlation-based guided filtering in shape from focus[J]. Microscopy Research and Technique, 2021, 84(7): 1368–1374.

[11] Onogi S, Kawase T, Sugino T, et al. Investigation of shape-from-focus precision by texture frequency analysis[J]. Electronics, 2021, 10(16): 1870.

[12] Nomnob N, Kitjaidure Y. Adaptive window size multi-focus images fused based on Sum of Modified Laplacian[C]//The 4th Joint International Conference on Information and Communication Technology, Electronic and Electrical Engineering (JICTEE). IEEE, 2014: 1–4.

[13] Ngo T D, Kashani A, Imbalzano G, et al. Additive manufacturing (3D printing): A review of materials, methods, applications and challenges[J]. Composites Part B: Engineering, 2018, 143: 172–196.

Part 5

Production Philosophies and 4th Industrial Revolution

13 Best Practices for Mitigating Supply Chain Risks in Industrial IoT

*Lubna Luxmi Dhirani, Ali Akbar Shah,
Qasim Arain, Bhawani Shankar Chowdhry,
Abi Waqas, and Tanweer Hussain*

13.1 DIGITAL THREATS

In today's world, where manufacturing and fintech environments highly rely on cutting-edge technologies (i.e., IoT, digital twin, cloud, AI systems, etc.) and communication networks for processing, storing, and transmitting data, it is essential to develop a secure environment for protecting data. The prospects of the digital economy depend on the value created from data. Without data, there would be no analytics, adaptive learning, or future optics. Data used in supply chains may carry both sensitive/personal and non-sensitive information that is subject to Data Protection Regulations and Controls [1]. These regulations have been imposed to protect the confidentiality, integrity, and availability of data. However, sharing data online is risky and an individual must take preventive measures to remain secure and be not vulnerable to hackers. Mitigating cybersecurity risks is extremely important in today's digital world. With the increasing use of technology and the internet in our daily lives, cyber threats have become more complex and sophisticated. These threats can come in various forms and scale, such as insider threats, espionage, ransomware, denial of service (DoS), data breaches, and more.

Cyberattacks may have disastrous repercussions for people, companies, and even countries. These can lead to monetary losses, reputational harm, legal responsibilities, and even hazards to national security. Moreover, sensitive information such as bank information, medical data, and intellectual property might be compromised by cybersecurity flaws. The Kesaya cyberattack [2] raised global awareness regarding the data loss, financial and operational impact of supply chain attacks [3–5]. As per the statistics, the manufacturing cyberattacks are being increased by 60%. 452 Operation Technologies (OT) based vulnerabilities had been identified and more than 10 million people were affected as a result of supply chain attacks targeting 1,743 entities in 2022 [6]. The increasing number of cyberattacks and threat landscape needs to be mitigated in order to protect the production and process environment. As a result, it is essential to reduce cybersecurity risks by putting in

DOI: 10.1201/9781003376620-18

place strong security policies and measures including firewalls, anti-virus software, encryption, access restrictions, and staff training. Frequent vulnerability testing and security assessments may also aid in locating possible security holes and enhancing security posture.

This chapter describes cybersecurity challenges in the IIoT, clarifies the range of various security and communication standards and how alignment may be achieved and offers best practices to reduce cyber threats and standardization issues that arise in Industry 5.0/IIoT ecosystem.

13.2   CYBER ATTACKS IN INDUSTRIAL IOT (IIOT)

The Industrial Internet of Things (IIoT) has transformed the way organizations operate, enabling them to optimize processes, improve efficiency, and reduce costs. *"IIoT is a network of smaller, smart devices that connect to form systems that can collect, exchange and analyze data in real-time in settings like a warehouse or factory"* [7]. The actionable data collected from IoT in the supply chain powered by novel communications methods (i.e., 5 G/6 G, Wi-Fi 6/7, TSN Ethernet) [8] can enable manufacturers not only to meet the existing demands and challenges but also to prepare for the next one. IIoT currently is exposed to various privacy, security, ethical, and regulatory issues [1]. The whole vision of the digital and circular economy is based on the security and sustainability aspects of Industrial IoT. By 2030, there will be 17 billion IoT-connected devices [9], and the IIoT market is anticipated to touch the $1.3 trillion mark [10]. Considering this, with the increased adoption of IIoT devices, the cyber risks and threat landscape associated with the supply chain will escalate as well. This section focuses on the best practices for mitigating cyber risk across IIoT supply chains.

13.2.1   CYBER RISK

Cyber risk is defined as the potential damage caused by a threat exploiting an existing vulnerability. Vulnerabilities are flaws or weaknesses related to hardware, software, noncompliance, etc. (e.g., open software backdoor or active credentials related to a former employee, etc.). These flaws can turn into threats only if they are identified and exploited (e.g., hacker entering through the software backdoor and eavesdropping or misusing the credentials of a former employee affecting the confidentiality, integrity, and availability of data). This is when a malicious hacker has the potential to cause actual harm to an infrastructure. The impact of cyber risk depends on the assets compromised. If critical assets are breached, the supply chain environment may lead to loss of command and control and financial losses. In 2021, the U.S. Colonial Pipeline paid $4.4. million in cyber extortion [11], Saudi Aramco, an oil-based company in Saudi Arabia faced $50 million in cyber extortion over leaked data, and this was not the first time Aramco's security was breached [12]. In 2012, Saudi Aramco was largely hit by a self-replicating virus called Shamoon that infected 30,000 Windows-based machines inside the company; as a result, the company suffered weeks of downtime, that leads to operational, financial, and data loss [13]. To provide cyber resilience and protect the industrial environment, it is important to understand and evaluate the cyber risks in IIoT supply chain. This section demonstrates different types of cyberattacks that the IIoT is susceptible to [3,4].

13.2.1.1 Password Attack

Password attacks are carried out by hackers who have the ability and skills to crack the password using password-cracking software or hardware. There are different types of password attacks, such as credential surfing, keylogger attacks, dictionary attacks, and brute force attacks [14]. Such attacks can be prevented by using the following best practices:

- A Single Sign-on (SSO) solution enables a single secured access point for authenticating access to different systems used within the manufacturing environment. Critical Industrial Control Systems are often isolated or micro-segmented for reducing cyber risks, each of these systems in the supply chain may have different credentials for access management. This increases the time and effort to manage different passwords across different devices, especially when employees forget the credentials and require a new one. An SSO mitigates these issues.
- Using multi-factor authentication (e.g., authenticator app) is considered more secure than an SSO, as it uses two different methods for authentication of a user's identity. This may include inherence (i.e., biometric verification), knowledge (i.e., password or answering a question and possession (device or a security token).
- SSO and MFA have limitations, when it comes to legacy equipment (secure shell (ssh), remote desktop servers, database servers, filesharing servers, custom web apps on web servers, etc.), as such security mechanisms were not built into them. In such scenarios, solutions such as Twingate may be beneficial as it allows to layer on MFA to any legacy equipment [15].
- It is also recommended to avoid using the same passwords for logging on to different websites and devices.
- Enforcing strong security, Identity, and Access Management policies and controls.
- Using security tools such as Security Information and Event Management (SIEM), Security Orchestration, Automation and Response (SOAR), and Extended Detection and Response (XDR) for building threat intelligence and incidence response [16].
- Deploying a robust anti-virus tool.
- Using strong passwords including special characters, alphanumerical, and digits.

13.2.1.2 Malware Attack

In 2022, the manufacturing sector *"suffered more than 437 ransomware (malware-based) attacks, making up more than 70% of these types of costly and disruptive assaults that industrial companies faced last year"* [17]. Looking at the operational disruption and financial impact of these attacks, it is essential to understand how these attacks are launched and find ways to mitigate them.

Malware are referred to as viruses like trojans, ransomware, worm, spyware, etc. Around 560,000 new pieces of malware are detected on daily basis. Having more than 1 billion malware programs in the forms of adware, spyware, etc. easily available to malicious actors on the dark web for as low as $13, the impact and threats that they expose to the Industrial IoT environment are massive. [18] As these viruses are being released every day, even the best anti-virus tools may fail to identify the viruses and quarantine them. Once, a malicious actor gains access to the command and control (C&C), they have full control of the environment and can exploit the confidentiality, integrity, and availability of data. These attacks can be carried out by identifying vulnerabilities within the hardware, software, or network environment, e.g., using credentials of a former employee, malicious insider compromising air-gapped systems by inserting an infected USB, lack of security features and updates in legacy systems, etc. [19]. To mitigate these types of attacks, the following security measures and best practices must be integrated within the supply chain environment:

- Incorporating data diode for securing data transfers. *"Data diodes are used to segment and defend networks and transfer information in one direction. They allow data to be sent from a secured network/segment to external systems and users (e.g., the cloud, a remote monitoring facility, regulatory bodies), without creating a threat vector back into the secured network"* [20,21]. As data diodes are hardware-based entities, carrying out an online attack on it is impossible. A data diode is viable for securing air-gapped legacy equipment as it is uni-directional but also benefits from remote connectivity.
- Incorporating a Demilitarized Zone (DMZ) strategy would add an extra layer of security. DMZs are usually configured with a firewall that enables the manufacturing environment to filter the malicious traffic. It acts as an active defense mechanism for critical business applications and servers, as at any point if the environment's internal network is breached, the critical business processes suffer no damage as they remain in the DMZ.
- It is recommended to regularly patch the systems mitigating the potential software/OS vulnerabilities.
- It is essential to have intrusion detection and prevention systems (IDPS). IDPS is considered an effective method for defending the novel cyber threats in ICS, that go undetected otherwise. *"An intrusion detection system (IDS) is a tool that helps to detect intrusions by inspecting the network traffic,"* [22] whereas *"an intrusion prevention system (IPS) is a network security tool (which can be a hardware device or software) that continuously monitors a network for malicious activity and takes action to prevent it, including reporting, blocking, or dropping it, when it does occur"* [23].

- The environment must be protected by internal and external software/ hardware-based firewalls.
- Besides the previous, human error remains the biggest threat vector and can only be mitigated by developing a cyber awareness environment. Regular training could be beneficial and the course work must be updated on the novel as well as advanced security risks related to connecting unauthorized devices in the network, clicking on external links that could potentially be malicious, access management, etc.

Watering Hole attacks are categorized under malware-based attacks. These types of attacks are carried out by malicious actors hacking a website that is frequently used by a specific group of audience. The hackers profile their victims online before targeting them. Once, the website the victims use is compromised, then that website can be used for distributing malware or manipulating user information. Watering hole attacks and supply chain attacks can look similar but they are slightly different. A supply chain attack delivers malware through the weakest element in an organization's network, such as a supplier, vendor, or partner, e.g., Kesaya cyberattack [19]. Whereas SolarWinds (2020), Holy Water (2019), OceanLotus (2018), NotPetya, CCleaner, and Montreal-based International Civil Aviation Organization attacks (2017) were all watering hole attacks, that affected millions of people globally [24]. Watering hole attacks can be mitigated by using the following best practices:

- Developing a culture of secure browsing amongst the users. By practicing good cyber hygiene (i.e., using a good anti-virus and IDPS tool, firewalls, patching and updating software on regular basis, strong passwords, incorporating Zero Trust (ZT), multi-factor authentication, device encryption, network security, having a backup and disaster recovery strategy) can protect the environment from Watering hole based attacks.
- Industrial IoT is data-driven therefore it is essential for it to have security and privacy regulations and controls incorporated within the environ- ment's cybersecurity strategy to mitigate potential cyber threats [1].
- Data is the most valuable entity in the Industrial IoT environment.
- Checking the authenticity of web links. A new initiative by CyberSkills Ireland CheckMyLink supports safer online browsing and provides information on whether the website is a scam, phishing, or legitimate [25]. This could be a good approach for the ICS environment to follow and protect from falling prey to scams and phishing attacks.

13.2.1.3   Man-in-the-Middle Attack (MITM)

The MITM attack is a passive attack, carried out by malicious actors with the intent to intercept/eavesdrop on an ongoing communication. These types of attacks undermine the confidentiality and integrity of data/communications and in doing so, they jeopardize data or devices for malicious exploitation. Experts estimate that around 35% of attacks that exploit cyber vulnerabilities have been Man-in-the-Middle attacks [26]. The impact of MITM attacks is widely known because of the serious implications they may cause. If a MITM attack is successfully carried out at the factory control network level, it may lead to Programmable Logic Controller's (PLCs) loss of command and control, safety, security, and data breaches [27]. As discussed in section 13.1, considering a digitally transformed environment these attacks tend to disrupt the environment and such threats need to be mitigated for building cyber resiliency in Industrial IoT.

MITM attacks can be mitigated by using the following best practices:

- *"Using deep packet inspection (DPI) and deep flow inspection (DFI) during network monitoring. DPI and DFI provide network monitors with information such as packet length and size. They can be used to identify anomalous network traffic"* [28].
- Incorporating Virtual Private Networks (VPNs) for data sharing prevents hackers from infiltrating as the data sent via VPN is encrypted.
- Using (https) secured websites makes the job complex for a malicious actor to obtain a valid certificate for a domain that is not managed by them and hence prevent eavesdropping. In situations of stolen credentials, multi-factor authentication acts as a second method for verifications and thus prevents malicious actors from gaining access and launching a MITM attack.
- Avoid using public, unsecured, and free Wi-Fi networks or connections.

13.2.1.4 Phishing Attack

One of the most prevalent and pervasive types of cyberattacks is phishing. It is a form of social engineering attack in which the hacker poses as a reliable contact and sends the victim fraudulent emails [29]. Unaware of this, the victim opens the email and either opens the attachment or clicks on the malicious link. Hackers can access private data and login passwords this way or even launch malware/keylogger attacks [30]. The HSE Ireland became a victim of Conti and suffered a phishing attack in May 2021, the attack soon escalated to a ransomware level encrypting all healthcare records, and blocking HSE's access to them. Without sufficient incident response and backup strategy, this issue turned into a national crisis [31]. In the past few years, credential harvesting has been of top interest to malicious actors, using different social engineering techniques for stealing login credentials and selling them on the dark web for financial benefits. The more details/personal information/a certain record from a breach/harvesting, the higher value is sought on the dark web. Manufacturing industries such as Pharmaceuticals and Medical devices, Genomic Data Centers, etc. deal with sensitive, personally identifiable information and healthcare data. The impact of these data breaches can lead to targeting/profiling or harming specific people having certain diseases, or racial or ethnicity-based target crimes. With the escalating geopolitical risks, the impact of such threats is yet to be assessed.

The best practices for mitigating phishing-based attacks [29,30] are:

- Creating a cyber awareness environment. Regularly upskilling, training employees, and running phishing simulation tests for building awareness.
- Having an incident response and backup strategy enables cyber resilience.
- Having strong controls implemented prevents unauthorized access and usage of unsecured sites.
- Using a robust firewall and keeping the systems patched.
- Using email signature certificates.
- Imbibing the "think before you click" approach by examining an email before clicking on any link.
- Enabling Multi-Factor Authentication (MFA) on devices.

13.2.1.5   Denial of Service Attack (DoS)

The most significant threat to Industrial IoT is a DoS attack in which malicious actors target the network infrastructure and the physical devices (i.e., servers, workstations, routers, switches, etc.). The physical devices receive countless incoming requests that ultimately slow physical devices or shut them down, leaving the authorized service request unattended [32]. When a DoS attack is carried out on connected multiple systems it is called a Distributed DoS (DDoS) attack [33,34]. Such attacks can be mitigated by undertaking the following steps:

- Testing the environment's risk appetite and security assessment.
  - Network vulnerability audits help in identifying existing weaknesses/ flaws within an industrial environment, that may have been overlooked.
- Developing a multi-level strategy for intrusion detection, prevention, and threat management (i.e., firewalls, a virtual private network (VPN), network and micro-segmentation, load-balancing, etc.).
- Developing cyber resilience by using multiple servers and cloud security.
- Having an incident response in place.
- Implementing best practices and aligning them with the industry's cybersecurity strategy.

13.2.1.6   Structured Query Language (SQL) Injection Attack

SQL injection (SQLi) attacks are a type of cyberattack that target the security vulnerabilities in web applications that rely on SQL databases to store and retrieve data. The attack exploits the vulnerability by injecting malicious SQL code into the database, which can result in data theft, data manipulation, and unauthorized access to sensitive information [35]. SQLi ranks among the top 3 web application security risks and is categorized into three types, based on the methods for accessing backend data and the impacts [36].

- *In-band SQLi (Classic):* In-band SQLi attacks are the most common type of SQL injection attacks. They occur when an attacker uses the same communication channel to send the malicious SQL code and receive the results. This can be done either through error messages or through the results of a query. In-band SQLi attacks are the most damaging because they can result in complete control of the database by the attacker.

- *Inferential SQLi (Blind):* Inferential SQLi attacks, also known as Blind SQLi, are more difficult to execute than In-band SQLi attacks because they require a different communication channel for the attacker to receive the results of the malicious SQL code. Inferential SQLi attacks are characterized by the fact that the attacker does not receive the results of the query directly but must infer them by observing the behavior of the system. This can be done through a series of true/false questions or time delays in the response of the system. Blind SQLi attacks are less damaging than In-band SQLi attacks but can still result in unauthorized access to sensitive information.

- *Out-of-band SQLi:* Out-of-band SQLi attacks occur when an attacker uses a different communication channel to send the malicious SQL code and receive the results. This can be done through DNS requests or HTTP requests. Out-of-band SQLi attacks are the least common type of SQL injection attacks but can still be damaging if the attacker can access sensitive information.

By classifying SQL injections based on their methods of execution and potential damage, security professionals can develop appropriate countermeasures to prevent these attacks from occurring. To prevent such types of malicious attacks, the following best practices would be helpful: [37]:

- Using stored procedures that are pre-compiled and parameterized.
- Implementing prepared statements can also be used for parameterizing SQL queries and prevent SQLi attacks.
- Using Object Relational Mapping (ORM) Frameworks (i.e., Hibernate, Entity Framework, and Django) for providing abstraction layers between the application and database. These frameworks would automatically generate parameterized SQL queries, reducing the risk of SQLi attacks.
- Applying access controls, employing zero trust and least privileges for mitigating unauthorized access to database and web interfaces.
- Validating user input on both the client and server side. This includes checking the input data type, length, format, and range. Validating user input, also ensures that verified/compliant data is sent to the database.
- Employing vulnerability scanners (i.e., SQLmap or Netsparker) for identifying SQLi vulnerabilities in the application.

While Web Application Firewalls (WAP) were traditionally used, modern technology is moving toward run-time application self-protection (RASP). WAP implemented at the network layer, provides the first lines of defense, detects and filters out potential threats before they could get to the target applications [38]. Whereas RASP enables customizing security for each application running in the environment, identify vulnerability at the software's application layer and address them in real time [39]. The authors recommend making this choice of implementing the former or latter based on the industry's operational environment and cybersecurity strategy. Although the SQLi attacks are happening at the web application levels, they still have the ability to affect a fully connected IIoT supply chain environment. Therefore, it is essential to look at all the security gaps in enabling technologies used in the smart factory, that could potentially lead to operational downtimes and impact cyber resilience.

13.2.1.7 Insider Threats

- These types of attacks are carried out by an existing employee who may have malicious intent. The motivation behind such attacks could possibly be rage, greed, or at times mere negligence. Such attacks are hard to anticipate [40] and the best practices for dealing with these threats are:
- Having awareness sessions designed for employees regarding network security.
- Employing zero trust methods for critical assets.
- Enforcing strong organizational policies and controls for mitigating unauthorized access and control.
- Employing Cybersecurity & Infrastructure Security Agency's (CISA) insider threat mitigating strategy in IIoT would also be helpful as it provides a roadmap for identifying, assessing, and managing insider threats [41].

13.2.1.8 Zero-Day Exploit (ZDE)

"A zero-day (0 day) *attack takes place when hackers* exploit *the flaw before developers have a chance to address it"* [42]. Stuxnet has been the most famous example of 0 day, the virus that infected the PLC software In 2021, *"the 0 day PwnedPiper vulnerability impacted pneumatic tube systems used by hospitals to transport bloodwork, test samples, and medications. The attackers could exploit flaws in the control panel software, which allowed for unauthenticated and unencrypted firmware updates"* [43,44]. The best practices to mitigate these types of exploitations are:

- Building threat intelligence is essential for developing a proactive cybersecurity strategy.
- Alike the threat mentioned earlier, developing an incident response and effective cybersecurity strategy is important for enabling cyber resilience.
- Implementing strong security controls, policies, updated/patched systems, and using a robust anti-virus solution would enable mitigating the risks.
- An article on Imperva titled "Zero Day Exploit" suggests *"deploying a web application firewall (WAF) on the network edge that reviews all incoming traffic and filters out malicious inputs that might target security vulnerabilities."*

13.2.1.9 Cryptojacking

This type of attack is carried out when hackers can access one's computer for crypto-mining. This can be performed when a victim clicks a malicious link.

Unaware of the consequences, the mining code runs on the computer compromising the system resources [45]. Such attacks can be mitigated using the following best practices:

- Deploying an advanced network monitoring solution could be beneficial as that would provide insights/visibility on the network performance and ways to remediate issues.
- Using tools such as ad-blocker or anti-crypto miner extensions on web browsers.
- Using endpoint protection capable of detecting known crypto miners.
- Keeping the systems patched, mitigating potential vulnerability exploitation.
- Keeping tabs on resource usage, computational power, and overheating.

The European ENISA threat landscape 2022 reported the top 8 threats affecting the IIoT environment, that are: ransomware, malware, social engineering threats, threats against data, threats against availability (DoS), internet threats, misinformation, and supply chain attacks [46]. The authors have reviewed these threats with examples in section 13.1 and provided best methods to mitigate them. With the increasing capabilities of malicious actors, it has become essential to detect and fix vulnerabilities before they could be exploited at the hacker's hands. The next section addresses risks associated to the supply chain ecosystem.

13.3 SUPPLY CHAIN RISKS IN IIOT

As the integration of IIoT and digitalization in Supply Chain Management (SCM) deepens, the cybersecurity risks facing organizations continue to grow. With the increasing numbers of supply chain attacks, it is interesting to know that the key factors contributing to these risks include: increased connectivity, lack of insights, visibility and control, lack of standardization, cyber law, privacy, and regulations [1,8,47]. The supply chain cyber risks in IIoT encompass the potential hazards associated with the deployment of IIoT devices throughout the supply chain ecosystem. These risks can manifest in various forms, including data breaches, system failures, and targeted cyberattacks. The presence of IoT devices in the supply chain creates numerous entry points for attackers, making it challenging for organizations to safeguard their operations [8,48]. Furthermore, when multiple organizations participate in the supply chain, the associated risks are exacerbated, as each entity introduces its own set of vulnerabilities that can be exploited by cybercriminals.

The integration of IoT devices into supply chains is driven by their ability to enable real-time monitoring, automation, and data analysis. These features can significantly improve operational efficiency, reduce costs, and streamline processes [49]. However, the very benefits that IoT devices offer also create a complex web of interconnected systems, which increases the attack surface for potential cyber threats [8].

One of the critical factors contributing to the heightened risk in supply chain cyber risks in IIoT is the diversity of technologies, hardware, and communication protocols used across the ecosystem. This heterogeneity complicates efforts to implement consistent security measures, making it more challenging to protect the supply chain from cyber threats. Moreover, each organization in the supply chain may have its own set of security policies and practices, which may not align with those of other participating entities. This lack of standardization can result in security gaps and inconsistencies, leaving the entire supply chain vulnerable to attack [1,8,50].

Additionally, the increased connectivity and interdependence between various organizations in a supply chain can lead to a domino effect in the event of a cyberattack. If a single organization in the supply chain is compromised, the effects can quickly propagate through the entire network, affecting multiple parties and causing widespread disruption [51]. This ripple effect can be further amplified when critical components or services are shared by several organizations within the supply chain.

By considering the aforementioned information, it can be concluded that the supply chain cyber risks in IIoT are a growing concern due to the increasing adoption of IoT devices, the complexities introduced by multiple organizations within the supply chain, and the inherent vulnerabilities arising from the lack of standardization and interconnected nature of the ecosystem. To mitigate these risks, organizations must adopt a proactive and collaborative approach, focusing on risk assessment, security-by-design, vendor management, and continuous monitoring to ensure the security and resilience of their supply chain operations [1,8,52].

13.4 BEST PRACTICES FOR MITIGATING CYBER RISK ACROSS IIOT SUPPLY CHAIN

Incorporating enabling technologies in the supply chain environment exposes the IIoT susceptible to various risks and threats, many of which have been discussed throughout this chapter. To mitigate a manufacturing environment from the mentioned risks, it is essential to stay informed of threat vectors, have threat intelligence built, and an actionable incident response plan in place. One must adapt the best practices, proactive and holistic approaches to cybersecurity, and learn from the breaches industries have suffered in the past. Incorporating standards, strong policies, and regulations would enable in developing cyber resilience in the infrastructure [1,8,46]. The authors provide their recommendations in this context that follows:

- Establishing a comprehensive security framework to align an industry's operational and cybersecurity strategy, standards, policy, and regulations for achieving a common cyber resilience goal.
- To stay ahead of potential cyber risks, an industry must conduct regular risk assessments across the supply chain. The NIST Risk management Framework [53] is one of the effective method for managing risks as it

provides "*a systematic process that integrates security, privacy, and cyber supply chain risk management activities into the system development life cycle. The risk-based approach to control selection and specification considers effectiveness, efficiency, and constraints due to applicable laws, directives, Executive Orders, policies, standards, or regulations. Managing organizational risk is paramount to effective information security and privacy programs; the RMF approach can be applied to new and legacy systems, any type of system or technology (e.g., IoT, control systems), and within any type of organization regardless of size or sector.*" In terms of information security and risk, the ISO 27001 [54] also provides a comprehensive approach for mitigating potential risks and security controls. As mentioned before each industry may have a different risk appetite and operations, so it is essential to implement a risk management framework that closely aligns with the industry and map it with relevant standards, controls and regulations. This would enable in identifying, assessing, managing and mitigating the potential risks the supply chain is susceptible to.

- Implementing zero trust and strong access management controls. A zero trust (ZT) strategy provides assurances that the resources accessed are securely irrespective of the location, adopts a least privilege strategy and enforces strong controls, and monitors and logs the network traffic [55]. Adoption of remote functionality and enabling technologies increases the threats landscape associated to human risks, compromising critical assets and resources. Implementing measures such as ZT and multi-factor authentication (MFA) would provide an additional layer of security. Although there are limitations related to legacy equipment to deploy these security features but with tools like Twingate, these features could be incorporated into them.

- Conducting audits is essential for ensuring that the security measures and controls applied are effective, it also provides assurance related to data governance and risk management [8].

- Ensuring End to End (E2E) data security is mandatory for IIoTs that deal with sensitive/personal identifiable information or healthcare data. The industry must ensure implementing effective data security controls for preventing data breaches. For examples, industries operating in the European jurisdiction must comply with General Data Protection Regulations (GDPR) [56]. Employing encryption methods (i.e., crypto-graphic protocols), using VPNs, anonymized or pseudonymized techniques are effective ways for securing data in transit, process, and at rest.

- Building threat intelligence is important for keeping tabs on novel and changing threat landscape. Using SIEM, SOAR tools or the MITRE ATT&CK ICS matrix [57] would enable in developing an effective incident response and build insights on the tactics, techniques, and procedures for mitigating threats.

- Ensuring all connected and IoT-based devices are up to date with the latest security patches for mitigating different vulnerabilities that the supply chain may be susceptible to. As the supply chains are susceptible to various risks, added lines of defense (i.e., robust firewalls, intrusion detection, and prevention systems) would be necessary.

- Regularly training employees on cybersecurity best practices and fostering a security-conscious culture is crucial for reducing the risk of insider threats and enhancing overall security posture.

13.5 CONCLUSION

As the proliferation of IIoT devices continues to revolutionize supply chains, industry's must remain vigilant in their efforts to mitigate cyber risks. In light of various topics covered in this book chapter, including digital threats, cyberattacks in industrial IoT, supply chain risks in IIoT, and mitigation strategies, the importance of a holistic approach to cybersecurity is evident. For building cyber resilience in digitally transformed supply chains, it is essential to detect, protect, and mitigate the novel threats that arise out of the complexities of connected devices, communication networks, processing and storage devices, lack of security mechanism in legacy systems, etc. There is no quick fix for building cyber resilience, each industrial environment may vary in terms of size, scope, operations, and functionality and would require different security standards, risk metrics, controls, and regulations. Additional layers of security would be required for bridging the gaps and aligning it with the industry's cybersecurity strategy. This chapter provides insights on ways for developing operational and cyber resilience in IIoT supply chain by employing the best practices and author recommendations.

REFERENCES

[1] L. L. Dhirani, N. Mukhtiar, B. S. Chowdhry, and T. Newe, "Ethical dilemmas and privacy issues in emerging technologies: A review," *Sensors*, vol. 23, no. 3, p. 1151, 2023.

[2] Ö. Aslan, S. S. Aktuğ, M. Ozkan-Okay, A. A. Yilmaz, and E. Akin, "A comprehensive review of cyber security vulnerabilities, threats, attacks, and solutions," *Electronics*, vol. 12, no. 6, p. 1333, 2023.

[3] M. Ghiasi, T. Niknam, Z. Wang, M. Mehrandezh, M. Dehghani, and N. Ghadimi, "A comprehensive review of cyber-attacks and defense mechanisms for improving security in smart grid energy systems: Past, present and future," *Electr. Power Syst. Res.*, vol. 215, p. 108975, 2023.

[4] K. Bitirgen and Ü. B. Filik, "A hybrid deep learning model for discrimination of physical disturbance and cyber-attack detection in smart grid," *Int. J. Crit. Infrastruct. Prot.*, vol. 40, p. 100582, 2023.

[5] F. Heiding, E. Süren, J. Olegård, and R. Lagerström, "Penetration testing of connected households," *Comput. Secur.*, vol. 126, p. 103067, 2023.

[6] CSO, "Supply chain attacks increased over 600% this year and companies are falling behind," 2022. https://www.csoonline.com/article/3677228/supply-chain-attacks-increased-over-600-this-year-and-companies-are-falling-behind.html.

[7] Ericsson, "How IoT in the supply chain can help manufacturers." https://www.ericsson.com/en/blog/2022/5/how-iot-in-the-supply-chain-can-help-manufacturers.

[8] L. L. Dhirani, E. Armstrong, and T. Newe, "Industrial IoT, cyber threats, and standards landscape: Evaluation and roadmap," *Sensors*, vol. 21, no. 11, p. 3901, 2021.

[9] 101 Blockchain, "IoT connectivity industry forecast By 2030," 2023. https://101blockchains.com/iot-connectivity-industry-forecast/#:~:text=As a matter of fact, and asset tracking and monitoring.

[10] IoTNow, "Industrial IoT market to reach US$1.3tn by 2032: Market expected to grow at 12.2% CAGR, says FMI." https://www.iot-now.com/2022/04/12/120692-industrial-iot-market-to-reach-us1-3tn-by-2032-market-expected-to-grow-at-12-2-cagr-says-fmi/.

[11] TechTarget, "FBI seized Colonial Pipeline ransom from DarkSide affiliate." https://www.techtarget.com/searchsecurity/news/252513302/FBI-seized-Colonial-Pipeline-ransom-from-DarkSide-affiliate#:~:text=The pipeline was shut down, of it had been recovered.

[12] CNBC, "Saudi Aramco facing $50 million cyber extortion over leaked data." https://www.cnbc.com/2021/07/22/saudi-aramco-facing-50m-cyber-extortion-over-leaked-data.html.

[13] T. and Francis, "The Cyber Attack on Saudi Aramco." https://www.tandfonline.com/doi/abs/10.1080/00396338.2013.784468?journalCode=tsur20#:~:text=On 15 August 2012%2C the,of its Windows- based machines.

[14] M. Mijwil, I. E. Salem, and M. M. Ismaeel, "The significance of machine learning and deep learning techniques in cybersecurity: A comprehensive review," *Iraqi J. Comput. Sci. Math.*, vol. 4, no. 1, pp. 87–101, 2023.

[15] Twingate, "Protect legacy technologies with multi-factor authentication." https://www.twingate.com/docs/protect-legacy-apps-with-multi-factor-authentication.

[16] Secureworks, "Understanding the difference between SOAR vs SIEM vs XDR." https://www.secureworks.com/blog/xdr-vs-soar-finding-the-right-tool-for-the-job.

[17] Engineering.com, "Manufacturing was the most targeted sector for ransomware attacks in 2022, says IBM." https://www.engineering.com/story/manufacturing-was-the-most-targeted-sector-for-ransomware-attacks-in-2022-says-ibm.

[18] DataProt, "A not-So-common cold: Malware statistics in 2023," 2023. https://dataprot.net/statistics/malware-statistics/.

[19] Logz.io, "How to defend your business against SQL injections," [Online]. Available: https://logz.io/blog/defend-against-sql-injections/.

[20] O. D. Defense, "Learn about data diodes." https://owlcyberdefense.com/learn-about-data-diodes/#:~:text=Data diodes are used to,back into the secured network.

[21] I. Cyber, "Implementation of data diodes can boost cybersecurity architecture at critical infrastructure installations." https://industrialcyber.co/analysis/implementation-of-data-diodes-can-boost-cybersecurity-architecture-at-critical-infrastructure-installations/.

[22] P. Vanin *et al.*, "A study of network intrusion detection systems using artificial intelligence/machine learning," *Appl. Sci.*, vol. 12, no. 22, p. 11752, 2022.

[23] Vmware, "What is an intrusion prevention system?" https://www.vmware.com/topics/glossary/content/intrusion-prevention-system.html#:~:text=What is an intrusion prevention,it%2C when it does occur.

[24] Malware Bytes, "Watering hole attack." https://www.malwarebytes.com/watering-hole-attack#examples.

[25] Cyberskills, "Check a website." https://check.cyberskills.ie/.

[26] VPN Overview, "Man-in-the-middle attacks: Everything you need to know." https://vpnoverview.com/internet-safety/cybercrime/man-in-the-middle-attacks/#:~:text=Real-World Examples of MITM,and make a quick score.

[27] H. Lan, X. Zhu, J. Sun, and S. Li, "Traffic data classification to detect man-in-the-middle attacks in industrial control system," in *2019 6th International Conference on Dependable Systems and Their Applications (DSA)*, 2020, pp. 430–434.

[28] Spiceworks, "What Is a man-in-the-middle attack? Definition, detection, and prevention best practices for 2022." https://www.spiceworks.com/it-security/data-security/articles/man-in-the-middle-attack/#:~:text=Man-in-the-middle attacks can also be detected,to identify anomalous network traffic.

[29] Y. Y. Lee, C. L. Gan, and T. W. Liew, "Thwarting instant messaging phishing attacks: The role of self-efficacy and the mediating effect of attitude towards online sharing of personal information," *Int. J. Environ. Res. Public Health*, vol. 20, no. 4, p. 3514, 2023.

[30] T. Xu, K. Singh, and P. Rajivan, "Personalized persuasion: Quantifying suscepti-bility to information exploitation in spear-phishing attacks," *Appl. Ergon.*, vol. 108, p. 103908, 2023.

[31] pwc, *Conti Cyber Attacks on the HSE*.

[32] A. Hernandez-Suarez *et al.*, "ReinforSec: An automatic generator of synthetic malware samples and denial-of-service sttacks through reinforcement learning," *Sensors*, vol. 23, no. 3, p. 1231, 2023.

[33] T. H. H. Aldhyani and H. Alkahtani, "Cyber security for detecting distributed denial of service attacks in agriculture 4.0: Deep learning model," *Mathematics*, vol. 11, no. 1, p. 233, 2023.

[34] BYOS, "Denial-of-service (DoS) attack prevention: The definitive guide." https://www.byos.io/blog/denial-of-service-attack-prevention#:~:text=For this%2C it is essential, before they overwhelm your network.

[35] V. Abdullayev and A. S. Chauhan, "SQL injection attack: Quick view," *Mesopotamian J. CyberSecurity*, vol. 2023, pp. 30–34, 2023.

[36] Imperva, "SQL (Structured query language) injection." https://www.imperva.com/learn/application-security/sql-injection-sqli/#:~:text=SQL injections typically fall under,data and their damage potential.

[37] G. Verhulsdonck, J. L. Weible, S. Helser, and N. Hajduk, "Smart cities, playable cities, and cybersecurity: A systematic review," *Int. J. Human–Computer Interact.*, vol. 39, no. 2, pp. 378–390, 2023.

[38] C. Point, "What is a web application firewall (WAF)?" https://www.checkpoint.com/cyber-hub/cloud-security/what-is-application-security-appsec/rasp-vs-waf/#:~:text=WAF provides a first line,that slip by the WAF.

[39] TRACEABLE, "WAF vs. RASP: A comparison and guide to leveraging both." https://www.traceable.ai/blog-post/waf-vs-rasp-a-comparison-and-guide-to-leveraging-both.

[40] X. Li *et al.*, "A High Accuracy and Adaptive Anomaly Detection Model with Dual-Domain Graph Convolutional Network for Insider Threat Detection," *IEEE Trans. Inf. Forensics Secur.*, 2023.

[41] I. and S. A. Cybersecurity, "Insider threat mitigation." https://www.cisa.gov/topics/physical-security/insider-threat-mitigation.

[42] Kaspersky, "What is a zero-day attack? - Definition and explanation." https://usa.kaspersky.com/resource-center/definitions/zero-day-exploit#:~:text=One of the most famous,logic controller (PLC) software.

[43] TechTarget, "Zero-day exploits reached all-time high last year report finds." https://healthitsecurity.com/news/zero-day-exploits-reached-all-time-high-last-year-report-finds.

[44] N. Peppes, T. Alexakis, E. Adamopoulou, and K. Demestichas, "The effectiveness of zero-day attacks data samples generated via GANs on deep learning classifiers," *Sensors*, vol. 23, no. 2, p. 900, 2023.

[45] S. Ullah, T. Ahmad, R. Ahmad, and M. Aslam, "Prevention of cryptojacking attacks in business and FinTech applications," in *Handbook of Research on Cybersecurity Issues and Challenges for Business and FinTech Applications*, IGI Global, 2023, pp. 266–287.

[46] ENISA, *ENISA Threat Landscape 2022*, no. November. 2022.

[47] Z.-H. Sun, Z. Chen, S. Cao, and X. Ming, "Potential requirements and opportunities of blockchain-based industrial IoT in supply chain: A survey," *IEEE Trans. Comput. Soc. Syst.*, vol. 9, no. 5, pp. 1469–1483, 2021.

[48] P. Radanliev *et al.*, "Cyber risk at the edge: Current and future trends on cyber risk analytics and artificial intelligence in the industrial internet of things and industry 4.0 supply chains," *Cybersecurity*, vol. 3, no. 1, pp. 1–21, 2020.

[49] C. Adaros Boye, P. Kearney, and M. Josephs, "Cyber-risks in the industrial internet of things (IIoT): towards a method for continuous assessment," in *Information Security: 21st International Conference, ISC 2018, Guildford, UK, September 9–12, 2018, Proceedings 21*, 2018, pp. 502–519.

[50] M. D. Stojanović and J. D. Marković-Petrović, "Deep learning for cyber security risk assessment in IIoT systems," in *Encyclopedia of Data Science and Machine Learning*, IGI Global, 2023, pp. 2134–2146.

[51] R. J. Raimundo and A. T. Rosário, "Cybersecurity in the internet of things in industrial management," *Appl. Sci.*, vol. 12, no. 3, p. 1598, 2022.

[52] K. Tsiknas, D. Taketzis, K. Demertzis, and C. Skianis, "Cyber threats to industrial IoT: A survey on attacks and countermeasures," *IoT*, vol. 2, no. 1, pp. 163–186, 2021.

[53] NIST, "About the risk management framework (RMF): A comprehensive, flexible, risk-based approach." https://csrc.nist.gov/projects/risk-management/about-rmf.

[54] I. Standards, "ISO/IEC 27001 and related standards Information security management." https://www.iso.org/isoiec-27001-information-security.html.

[55] Z. A. Collier and J. Sarkis, "The zero trust supply chain: Managing supply chain risk in the absence of trust," *Int. J. Prod. Res.*, vol. 59, no. 11, pp. 3430–3445, 2021.

[56] GDPR.EU, "Complete guide to GDPR compliance." https://gdpr.eu/.

[57] M. Attack, "ICS matrix." https://attack.mitre.org/matrices/ics/.

14 Machine Intelligence for Agile Manufacturing

S. Katyara, L. L. Dhirani, P. Long, T. Morell,
F. O'Farell, C. Edmondson, and B. S. Chowdhry

14.1 INTRODUCTION

Over the past two centuries, the industry has undergone significant transformation due to advancements in technology, particularly the shift from mass production to mass customization. This shift is driven by customer preferences and fast prototyping, putting pressure on manufacturers to be more agile to stay competitive in the market. Employing modern machine intelligence tools such as cyber-physical systems, fast communication protocols, the Internet of Everything, and collaborative robots can bring agility to the manufacturing ecosystem (1). These tools enable manufacturers to create fully connected production systems for better coordination across their value chain, leading to active intercommunication with end-users. This helps to continually update and upgrade processes, quickly adapt to system dynamics, reduce manufacturing waste, and improve overall productivity (2). As shown in Figure 14.1 (2), the connected ecosystem allows customers to co-create their own product of choice, which in turn helps the manufacturer to continuously improve and optimize the product. This reduces risks and increases productivity.

The existing manufacturing industry, primarily small and medium-sized enterprises (SMEs), which account for about 90% of the market share globally, is still tied to legacy-based production that inherently lacks flexibility and connectivity (3). Their manufacturing strategy resembles a waterfall development model, which is linear and sequential, and doesn't incorporate feedback, making it not resilient to intermittent process variations until the final product is manufactured. On the contrary, the agile model supports multiple production cycles to run simultaneously for faster iteration, as shown in Figure 14.2(a) (4). Figure 14.2(a) demonstrates how the risk associated with overall product development and production is reduced under the agile model due to its distributed nature and ability to take multiple low-stake and high-value actions. The manufacturing industry can transform their production into agile processes by adopting the right tools and techniques. These tools and techniques should be chosen to ensure worker augmentation, dynamic adaptation, bottom-up communication, and faster iteration, which are explicitly or implicitly linked to machine intelligence.

DOI: 10.1201/9781003376620-19

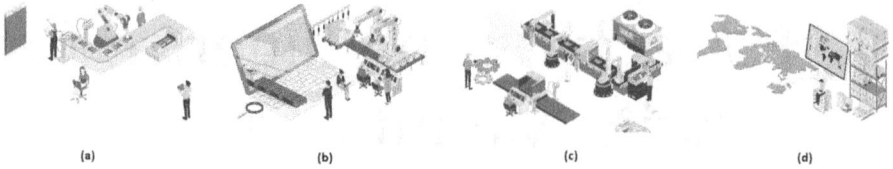

FIGURE 14.1 Life cycle of agile production system: (a) co-creation of product, (b) product optimization, (c) better process efficiency, (d) tracked product dispatch.

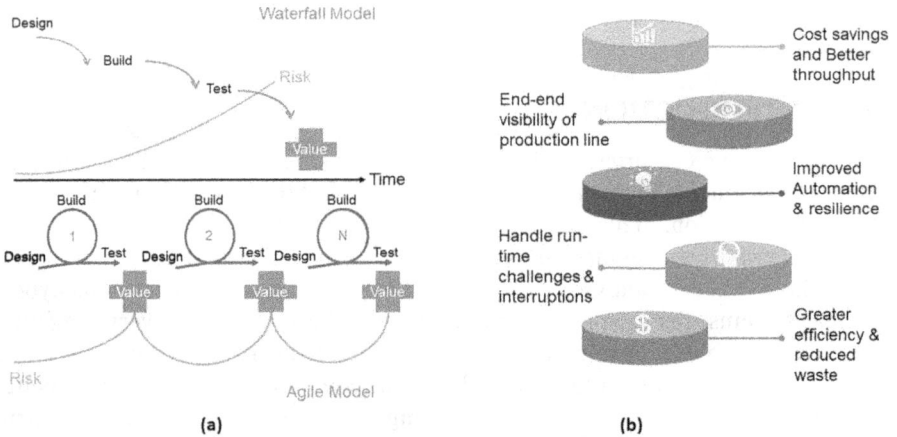

FIGURE 14.2 (a) Positives of applying machine intelligence into agile manufacturing conditions; (b) risk reduction from waterfall model to agile strategy in manufacturing.

Machine intelligence in agile manufacturing digitizes the entire workflow using big data analysis techniques to enhance operational efficiency, product quality, machine uptime, and cost-effectiveness. It gives the agile manufacturing process self-healing power to continuously learn, adapt, and improve under volatile conditions. The advancements in machine intelligence for agile manufacturing are summarized in Figure 14.2(b). While these machine intelligence tools enhance the performance, potential, and productivity of manufacturing processes, they also pose certain threats in the form of safety and ethics. Specifically, the use of collaborative robots and machine learning methods may compromise the trustworthiness and dependability of the human-centric workspace if proper cybersecurity plans, system safety standards, and ethical guidelines are not defined and followed (3).

This chapter examines the fundamental role, applications, and implications of machine intelligence in agile manufacturing. The evolution of the industry from mass production to mass customization is discussed in Section 14.2. Section 14.3 explores the use of autonomous robots to tackle challenges and opportunities related to mass customization. Section 14.4 covers the most common and effective key performance indicators (KPIs) and metrics used in agile production. Section 14.5 provides an overview of relevant safety standards and ethical protocols for the augmented workspace, and Section 14.6 analyzes and proposes the automation of feasible manufacturing tasks using the ACROBA ecosystem.

14.2 DEMYSTIFYING INDUSTRY 5.0

To date, we have experienced five industrial revolutions that have primarily been driven by technological advancements, resource availability, and socio-economic growth. as shown in Figure 14.3. The first industrial revolution began in the UK in 1740 and spread to other parts of Europe over the next 80 years. It was marked by the transformation of manual labor into machine-based manufacturing through the use of the spinning jenny, steam engines, and power looms. The second revolution, which took place in the late 19th century, was characterized by the use of electricity and the internal combustion engine. Significant innovations during this period included electric generators, telephones, and telegraphs, which greatly impacted transportation, communication, and the manufacturing industry. The third revolution, which lasted from 1960 to 2010, saw the integration of electronics and information technology to automate manufacturing processes. Key advancements during this period included the computer, barcode, and the internet, which led to the globalization of the economy and increased productivity through automation. Over the past decade, the manufacturing industry has undergone a transformation into the connected factories of the future, driven by the fourth revolution and the exploitation of technologies such as artificial intelligence, robotics, the internet of things, and cloud computing. This revolution is changing the way goods and services are developed and delivered, shifting the manufacturing industry from mass production to customization through digital tools and a knowledge-based workforce. The fifth industrial revolution is a continuation of Industry 4.0, focusing on flexibility, system resilience, and human alignment. It involves the integration of physical, digital, and biological systems through the use of biotechnology, nanotechnology, robotics, quantum computing, and other emerging technologies,

FIGURE 14.3 Evolution of industry from mass production to mass customization.

with the aim of creating new industries, products/services, and ways of living and working (5).

The outbreak of COVID-19 has significantly impacted the workstyle and manufacturing activities. As a result, lights-out operations and collaborative conditions have gained prominence to enhance the resilience of the system against intrinsic and extrinsic uncertainties, such as market volatility, supply chain uncertainties, and changing customer preferences. Industry 5.0 places emphasis on human-centric solutions that improve the ergonomics of operators by augmenting their workspace with digital tools. The three main objectives of Industry 5.0 are flexibility, resilience, and human alignment, as depicted in Figure 14.4. Flexibility allows the industry to adapt to changing conditions and exploit resilient solutions. Human alignment enables workers to evolve their work through solutions that are dependable and trustworthy. Industry 5.0 is an extension of Industry 4.0, which has established the foundation for human-machine and machine-machine collaboration through cyber-physical systems. The cyber-physical system connects the plant, logistics, supply chain management, and end-users under a common communication channel.

The ultimate goal of Industry 5.0 is to integrate the creativity of humans with the precision and strength of machines to achieve complex mass customization goals at a faster rate. Industry 5.0, along with Industry 4.0, provides a roadmap for the manufacturing industry to enhance productivity and system security. This includes the application of cutting-edge technologies such as artificial intelligence, robotics, the internet of things, and cloud computing.

- Machine Learning (ML): ML plays a crucial role in the manufacturing industry. It leverages the vast amount of data generated within connected systems to improve various aspects of the production process. For instance, it optimizes process scheduling, predicts maintenance needs, ensures product quality through analysis of irregularities, and supports informed

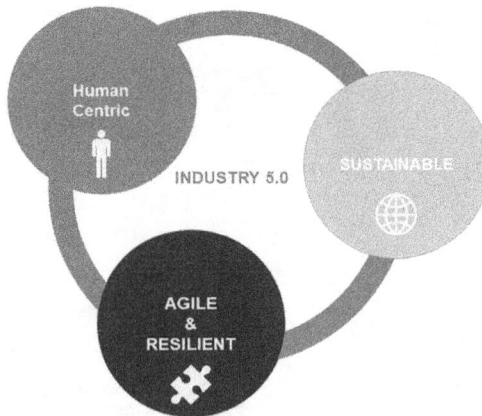

FIGURE 14.4 Three pillars of Industry 5.0.

decision making. By integrating ML, the industry aims to increase productivity and system security.

- Internet of Things (IoT): The IoT technology brings together human-to-machine and machine-assisted human-to-human communication, in addition to machine-to-machine interactions. It supports general intelligence and improved cognition throughout the interconnected manufacturing ecosystem. By connecting people, processes, things, and data into a cohesive loop, manufacturers are able to make context-aware decisions regarding production and dispatch.

- Collaborative Robots (Cobots): Cobots, with their small size and increased sensitivity, are able to work alongside human operators to complete tasks that require high levels of cognition and decision making as well as high precision in product assembly. Cobots are active and intelligent, allowing humans to guide them in learning new tasks that can then be generalized to different production conditions. These robots are designed to be safe, using either touch sensors or external perception feedback to detect their surroundings.

- Digital Twin: The use of digital twins in the manufacturing industry has a significant impact on designing, developing, and testing products using modern tools and techniques without affecting or altering the actual production line before feasibility is proven. Such feasibility studies for production scenarios are possible using their virtual replica, which reduces the risk of catastrophic failure in the event of a product cycle malfunction. This allows manufacturers to understand which solutions to adopt and how they will impact their setup when they are adopted, through sim2real transfer.

- Block Chain: Blockchain in the interconnected manufacturing ecosystem helps to increase transparency and trust throughout the entire industrial value chain, from sourcing raw materials to delivering the final product. It improves the efficiency of product design, engineering, production, and scaling in terms of time and cost.

- 6 G Connectivity: The rollout of 6 G connectivity helps reduce communication latencies related to process flows and enables the transfer of large amounts of data at faster bit rates over long distances. This not only improves the throughput of the manufacturing system but also aids in its resilience and response time against volatile customization conditions. It is a key enabler for improving the agility of the manufacturing system by iterating the production cycle at a faster rate.

- Metaverse: The Metaverse is a mixed reality experience that enables operators to monitor and inspect manufacturing systems through telepresence. With the help of 6 G connectivity and virtual-augmented reality settings, it puts humans in the loop by allowing them to take actions remotely based on multi-sensory feedback from exteroceptive sensors deployed at strategic locations. This allows experts to be present at multiple manufacturing sites simultaneously, making decisions about product designs, commissioning, overhauling, monitoring production processes, and more.

• Quantum Computing: Quantum computing enables faster optimization of products and production processes, both in terms of fabrication and raw material selection. Its ability to represent multiple options simultaneously is expected to aid manufacturers in discovering new raw materials and reducing time-to-market through optimized supply chain management.

It has long been debated that intelligent machines will replace human workers and take their jobs, which is partially true for tasks that involve tedious repetition, unnecessary cognitive burden, heavy physical labor, and are boring. However, it is also likely to create more skilled jobs for individuals. This transformation is expected to generate more employment opportunities and allow workers to develop new, relevant skills in emerging fields such as robotic process automation, data warehousing, and high-fidelity remote immersive operations. To ensure a smooth transition, continuous training, and workforce development initiatives will be necessary to bridge the gap between the available workforce and evolving market needs.

14.3 SMART AUTOMATION FOR MANUFACTURING

Agile manufacturers prioritize deploying autonomous solutions capable of handling task and domain variability in order to remain resilient against a rapidly changing environment driven by shifting user preferences. However, only about 16% of SMEs around the world use autonomous robots to drive their solutions toward mass customization, flexible supply chains, and (6). This lack of interest is primarily due to financial and resource constraints, as well as a lack of in-house expertise, awareness, and technological standardization. As a result, there is a strong demand for robust, adaptive, intelligent, and standardized solutions for SMEs in manufacturing. The main challenge is to design and deploy a general-purpose solution that is not only economically and technologically advanced enough to handle complex production processes, but also flexible, robust, and human-centric. Existing solutions are either pre-programmed (classical engineering) or generate optimal results only under controlled conditions (reinforcement learning agents), as shown in Figure 14.5. Additionally, these solutions often prioritize optimality and end-to-end behavior over safety, scalability, and feasibility, which are crucial for agile industrial tasks. Therefore, any smart solution deployed for agile production must ensure following:

• Modularity: Designing functional cells in terms of modules and (re-) arranging them according to the needs of scenarios (e.g., factory machines, robot platforms, perceptual systems, etc).

• Reconfigurability: Ability to quickly and easily change as per tasks requirements. However, it does imply complete reconstructing though but some relevant components (e.g., EOATs, fixtures, machine settings, etc).

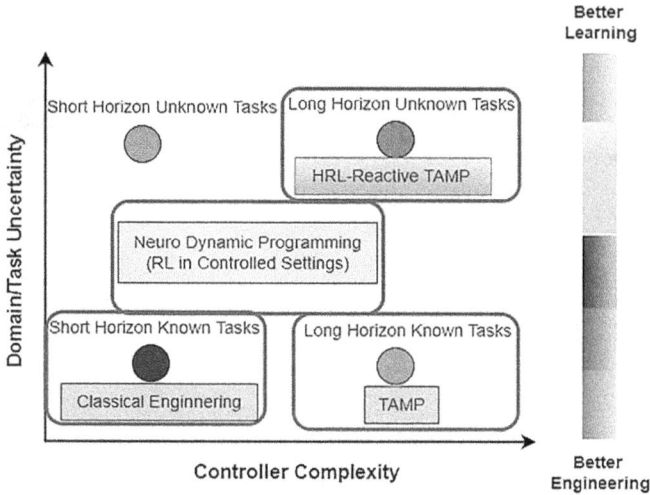

FIGURE 14.5 Algorithm exploitation against optimality and generalization.

- Programmability: Ease of programming is essential to ensure less deployment time and faster test iteration. Hence, changing application is just a matter of context by using right programming tools (e.g., Middleware, OS libraries, DRL frameworks, etc).
- Accessibility: Using common tools and techniques which are easy to handle and configure. This either does not require specialist or involving experts to operate and make modifications (e.g., mobile robots, 3D printers, RGBD sensors, etc).
- Safety: Development and exploitation should adhere to safe practices to avoid system mal-functioning and environmental hazards. This entails to employ additional safety sensors and protocols (e.g., Main PLCs, Laser Scanners, Motion capturing systems, etc.).

The journey toward agility entails capitalizing on modern digital techniques, especially the autonomous mobile robots (AMRs) that confer the service to improve operational efficiency, process speed, task precision, and workspace safety. AMRs use their onboard sensing, intelligence, and computing to understand their dynamic environment and perform actions to accomplish their goals effectively and reliably, as shown in Figure 14.6(a). The actions taken have associated effects which govern the affordance relation of mobile robots to their environments, i.e., change of state of environment, and the extent to which mobile robots are allowed to make decision independently depends upon their autonomy level. Both autonomy and affordance play important roles in the operation of mobile robots to exploit their navigation and manipulation capabilities at full capability within unstructured and flexible environments such as agile manufacturing.

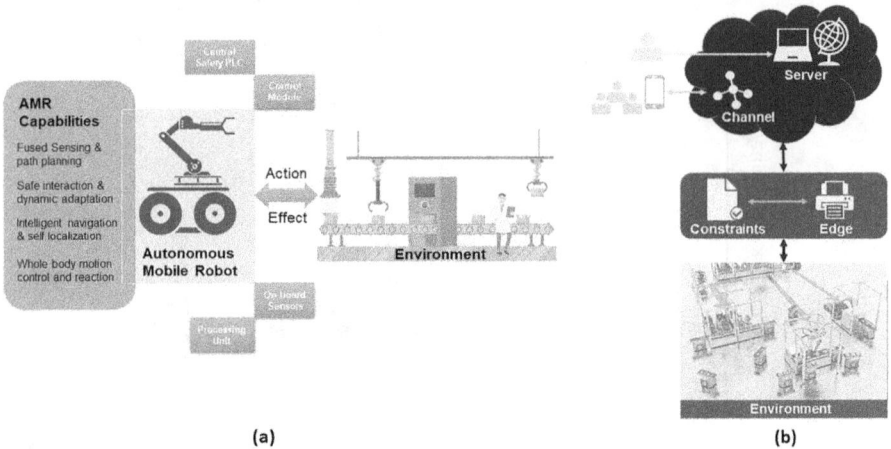

(a)   (b)

FIGURE 14.6 (a) How the application of autonomous mobile robots bring more flexibility and adaptability into manufacturing ecosystem; (b) conceptual design of active fleet management system.

Autonomous Mobile Robots (AMRs) facilitate effective and flexible material handling and intralogistics within the context of agile manufacturing, thereby addressing the issue of scarce skilled manual labor and enhancing productivity. The ease of reprogramming and reconfigurability of AMRs allows them to rapidly respond to market trends and minimize time-to-market. The integration of onboard sensors enables AMRs to collect and share data, facilitating the monitoring of the manufacturing ecosystem and supporting data-driven decision making. This results in an improvement in the end-to-end agility and competitiveness of connected production units. The ability of AMRs to operate continuously, 24/7, without fatigue or interruptions reduces the risk of accidents in the workspace and frees up human workers for other cognitively intensive tasks. This further contributes to overcoming the challenges of skilled labor shortage and turnover rates in manufacturing industry. Additionally, AMRs promote sustainability in manufacturing processes by reducing carbon footprints associated with manual handling and transportation of production materials and parts (7).

To fully utilize the capabilities of AMRs, multiple units are deployed in a factory. The fleet management system (FMS) is used to manage and control the AMRs. FMS is a combination of technologies, processes, and protocols that optimize AMR operations under complex conditions. It provides real-time location, status, and performance information to operators, allowing for informed decision making about resource optimization, path planning, and maintenance scheduling, reducing operational costs and increasing system agility. The basic FMS structure is depicted in Figure 14.6(b), consisting of a central server, user messaging through a channel, and constraints loaded by users. The central server deploys applications to the AMRs, and users have full control over all changes made, only proceeding after approvals.

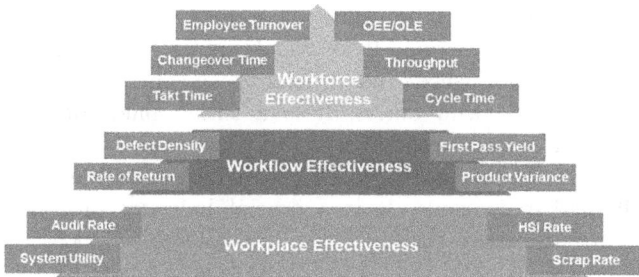

FIGURE 14.7 Metrics to measure and quantify the performance of agile manufacturing.

14.4 AGILE METRICS

To better understand and evaluate the performance of AMRs in agile production scenarios, a proper choice of task metrics and KPIs is essential. It enables to measure and improve the effectiveness of equipment and the workforce using dynamic planning and scheduling with AMRs. Although, defining customized or exploiting existing ones requires careful considerations to ensure that they comply with characteristics of interest, i.e., system operation, communication protocol, production efficiency, workspace dependability, etc. Overall, the metrics and KPIs for agile manufacturing are categorized into three groups, as shown in Figure 14.7.

14.4.1 WORKFORCE EFFECTIVENESS

These KPIs generally measure and evaluate the actions of an agent, whether human or robot, that improves the performance of the manufacturing system. These include:

- Employee Turnover: The indicator serves as a crucial measurement of not only the effectiveness and productivity of the team, but also its elevated value serves as a warning signal for manufacturing organizations to reflect on the potential negative impact on their productivity and operational efficiency. Monitoring this metric is imperative in ensuring the retention of skilled personnel and the attainment of organizational goals in a timely manner.
- Overall Equipment/Labor Effectiveness (OEE/OLE): It measures the extent to which the manufacturing machine or operator is productive and producing high-quality output as efficiently as possible with minimal downtime. However, it only measures the percentage of utilization compared to the maximum capacity. It also serves as an indicator for reducing resource waste by maximizing utilization.
- Changeover Time: It represents the average time required for a particular production line to switch from producing one product to another, demonstrating the flexibility and agility of the manufacturing ecosystem.

- Throughput: It measures the quantity of goods and services produced within a designated period. It is widely utilized in benchmarking to compare the performance and efficiency of one production line or manufacturing unit with others. It is a strong indicator of return on investment and system productivity.
- Takt Time: It gauges the speed at which the product is manufactured to fulfill market demands. It is a crucial indicator of time-to-market in agile manufacturing for high-mix, low-volume production and mass customization.
- Cycle Time: It is the overall duration required by a manufacturing system to produce a single unit of product. It does not only measure the efficiency of the system but also its ability to fulfill customized customer demands in terms of production.

14.4.2 WORKFLOW EFFECTIVENESS

The KPIs assess operational efficacy of production processes and the degree to which the productivity of overall manufacturing system is being improved. These include:

- Defect Density: Quantifies the proportion of defective units produced among the total volume of manufactured parts. If not monitored and addressed, it can negatively impact the reputation and profitability of the manufacturing company.
- First Pass Yield: Calculates the number of high-quality parts produced directly from the production process without any rework, in relation to the total number of parts produced. This key performance indicator reflects the stability of the production line and highlights any issues in the manufacturing unit.
- Rate of Return: Calculates the speed at which capital expenditures are recouped and the manufacturing unit begins to generate revenue. It considers unit costs and maintenance costs to determine the operating cost of the manufacturing plant. This indicator helps manufacturers understand whether their operations are profitable or unprofitable.
- Product Variance: Assesses the discrepancy between the actual physical product and its intended design as represented in the CAD model. This analysis provides valuable insight into the precision and accuracy of the manufacturing process and enables manufacturers to evaluate the conformity of the final product to customer specifications.

14.4.3 WORKPLACE EFFECTIVENESS

These KPIs determine the extent to which dependability, depreciation, and dispatch of the working environment are taken into account. These include:

- Audit Rate: Determines the frequency of safety and emergency protocol testing and verification in the workplace is determined by this analysis. It provides an assessment of the compliance of manufacturing operations with safety standards and highlights the need for regular verification to reduce the risk of accidents
- Health and Safety Incidence (HSI) Rate: Determines the incidence rate of injuries per 100 employees in the workplace over a 12-month period. This OSHA-certified metric serves to measure the occupational health and safety of workers in proximity to production processes
- System Utility: Evaluates the utilization rate of the existing production capacity of the manufacturing plant compared to its full potential. It incorporates information about downtime and operating hours of production units to determine the overall efficiency and effectiveness of the manufacturing facility.
- Scrap Rate: Assesses the efficiency of raw material utilization, including the proportion of discarded manufactured parts. Reducing this metric can improve the productivity of the manufacturing plant and mitigate challenges related to recycling or disposing of scrap materials.

14.5 SAFETY AND ETHICS

In the realm of agile manufacturing, safety, and ethics are two critical considerations that must be addressed to ensure the dependability and trustworthiness of the production process (8). To this end, it is imperative that industry-specific regulations and guidelines be strictly adhered to. The most widely used safety standards in the manufacturing industry are outlined in Figure 14.8. These standards are categorized into three classes based on the nature, size, and complexity of the production machines utilized. Type-A standards establish the fundamental safety features and principles that govern the operation of manufacturing machines, from design to full-scale implementation. To guarantee that production machines are in compliance with the acceptable level of risk tolerance, three iterative steps must be taken, including risk assessment, hazard identification, and the implementation of preventive measures. Notable standards in this category include ISO 12100 and IEC 61508. Type-B standards address either specific safety aspects, such as noise level, surface temperature, and safety distance (Type-B1), or the type of safeguards, such as light curtains, laser scanners, and PLCs, that can be utilized with a wide range of production machines. These standards serve as a front-end for Type-A standards, enabling the determination of potential hazards and providing concrete information on the steps to be taken to reduce risk. Some of the most commonly referenced standards in this category are ISO 13849-50-51 and IEC 62061. Type-C standards provide comprehensive guidelines and procedures for the safe installation and operation of specific machinery or complete machine setups, including industrial and collaborative autonomous robots. These standards take priority over Type-B and Type-C and define the modes of collaboration and levels of autonomy permitted for collaborative mobile robots in the workspace (9). Notable standards in this category include ISO 10218, TS 15066, ISO 3691–4, and EN 1525.

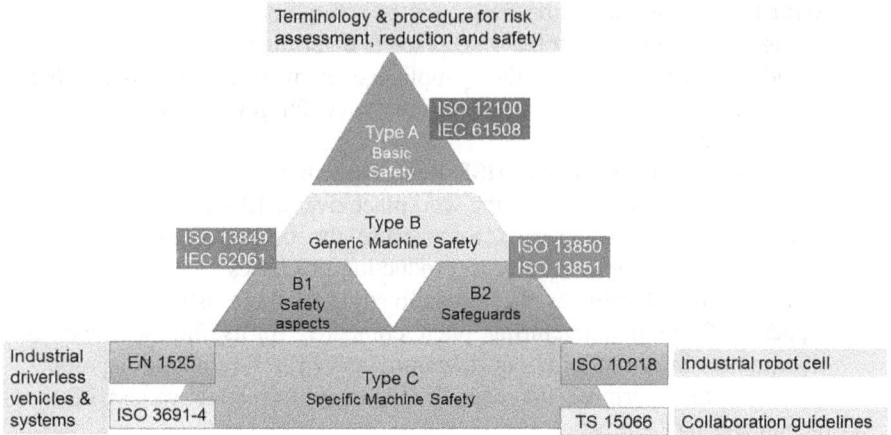

FIGURE 14.8 Commonly used safety standards in manufacturing industry.

The ethics of deploying Autonomous Mobile Robots (AMRs) in agile manufacturing settings is governed by three laws, which are depicted in Figure 14.9. These laws mandate that AMRs adhere to ethical principles such as avoiding collisions with obstacles, both static and moving, executing tasks within defined boundaries, safeguarding privacy, and conforming to human-centered guidelines. The first law dictates that AMRs must obey the commands of human operators and must be programmed to not cause any harm or make conflicting decisions. This means that AMRs should not learn new tasks beyond what has been programmed by the operator to prevent conflict and unintended consequences. The second law requires that AMRs, with all efforts and attention, must complete the task without violating the first law. This implies that AMRs, once programmed, must exhibit high levels of repeatability and precision in executing the assigned task and should be able to generalize to different conditions without breaking the principles outlined in the first law. The third law mandates that while fulfilling task goals and conforming to human-centered environments, AMRs must not compromise their presence or integrity while

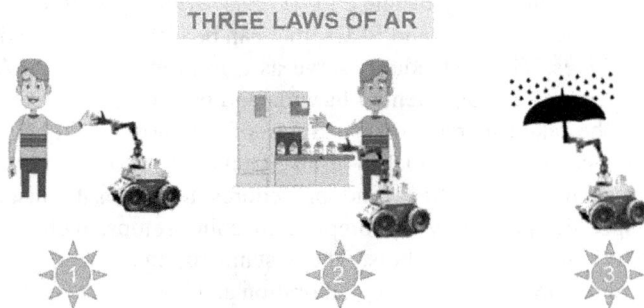

FIGURE 14.9 Three laws of autonomous robotics under manufacturing conditions.

adhering to the first and second laws. This means that AMRs should not take actions that protect humans but cause self-harm while accomplishing the task requirements.

14.6   USE CASES

The manufacturing industry encompasses a multitude of intricate processes and tasks aimed at achieving end product goals. Assembly and inspection are two of the most widely utilized tasks in this industry. The assembly task involves the sequential placement of parts in the desired position and orientation, adhering to task constraints to fulfill task requirements. This task is characterized by discrete actions aimed at achieving task goals. Whereas, the inspection task entails the analysis of the manufactured parts to determine if they conform to the design specifications and if any deviations exist, their nature and magnitude. These tasks are automated through the utilization of the ACROBA ecosystem, which constitutes a combination of reconfigurable hardware and software modules, aimed at automating the most frequently encountered agile manufacturing tasks. The present automated outlook of both assembly and inspection tasks performed through the ACROBA platform across agile cells is presented as follows:

14.6.1   ROBOTIZED AGILE ASSEMBLY

Bin picking is a widely used assembly task in industrial production processes. It involves sorting objects of varying shapes and sizes from cluttered bins using advanced perception techniques and motion planning algorithms to surpass human performance. ACROBA aims to automate bin picking by using a custom perceptual pipeline, including point-cloud segmentation, CAD matching, CNN-based localization, pose estimation, and motion planning with the MoveIt API. To evaluate ACROBA's bin picking performance, the following procedure has been implemented.

- Two trays are placed in the workspace, one filled with cubes and the other empty.

The trays are equipped with Aruco markers to be detected by the top-mounted camera, which allows the robot to move to the filled tray using the "move – to" skill. The onboard calibrated eye-in-hand camera then localizes the cubes using the perceptual pipeline, and the "pick" skill uses the pose information to grasp the cube without colliding with other objects or the tray. The robot then uses the "place" skill to bring the grasped cube to the empty tray and release it. This process is repeated for all cubes sequentially, as shown in Figure 14.10 using the industrial robot cell.

- The same procedure was also performed on a collaborative cell with varying lighting, as shown in Figure 14.11, to assess the robustness and adaptability of the ACROBA framework.

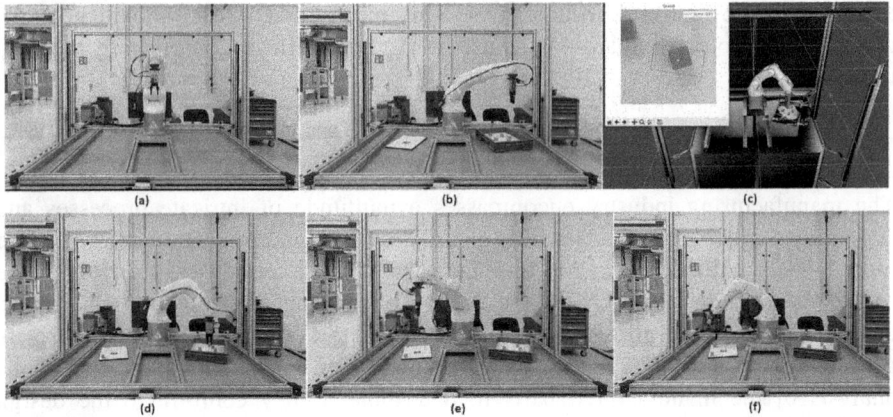

FIGURE 14.10 Industrial robot cell executing bin picking task on cubic objects under standard operating conditions: (a) initial environment configuration; (b) robot maneuvering to the filled tray; (c) eye-in-hand camera localizing the objects; (d) robot placing and releasing the object at the desired position in the empty tray.

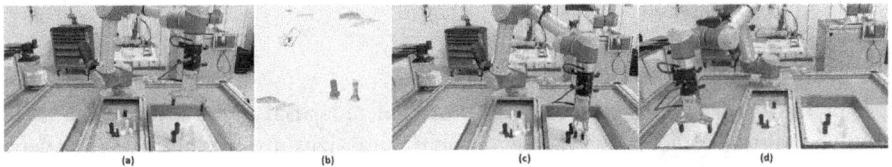

FIGURE 14.11 Collaborative cell performing bin picking task on SteriPack components under dynamic conditions: (a) robot detecting and localizing objects of interest in the tray; (b) object segmentation and pose estimation; (c) robot grasping the SteriPack knob under varying lighting conditions; and (d) robot transporting and placing the grasped component at the desired location in the correct position and orientation.

According to the Universal Robots (UR) report (10), 38% of the industrial workforce is engaged in manual bin picking, which is tedious, repetitive, and uninteresting for humans. This physical and psychological strain on workers' health highlights the need for automation. Automation not only improves productivity, reduces downtime, and ensures high quality, but also allows human workers to focus on more cognitive and delicate tasks. Despite the high demand and potential for automation, its slow adoption can be attributed to the required initial capital and expertise.

Most of the existing automated bin picking solutions are implemented by large OEMs. However, 69% of machine tending tasks are performed by SMEs, who face labor shortages (10). Despite their shortage of capital and expertise, SMEs are reluctant to adopt modern digitized solutions due to safety, trust, and complexity concerns. ACROBA platform provides a cost-effective, user-friendly solution designed specifically for SMEs, with minimal configuration time and usable by non-experts.

14.6.2 AUTOMATING INDUSTRIAL CO-INSPECTION

The inspection task aims to detect irregularities in manufactured parts to improve quality. Its success depends on the accuracy of the perceptual pipeline in reconstructing the 3D shape of objects. The ACROBA framework scans the parts, collects point-cloud information, reconstructs the 3D shapes, visualizes them as STL files, and matches them with standard CAD files to identify differences (irregularities). The co-inspection task involves collaboration between human and robot to achieve a common goal of faster and more reliable part inspections. The ACROBA platform uses certified sensors and adaptive controllers to monitor the human presence and regulate the robot's speed accordingly, ensuring safe human-robot coexistence and cooperation. This mode, known as speed and separation monitoring (SSM) according to TS 15066, involves defining working, warning, and danger zones so that the robot adjusts its speed when it detects a human approaching to prevent collisions (11). The strategy described later was implemented to automate part inspection while working closely with the human operator.

- For the purpose of inspection tests, a collaborative robot cell was employed. The robot, equipped with an eye-in-hand camera, executed a planar object scan using a pre-determined grid to acquire the required RGBD data for 3D reconstruction, as demonstrated in Figure 14.12. The SteriPack medical part was selected as a demonstration of the proof-of-concept and the results obtained were deemed satisfactory for an object of irregular shape. The 3D reconstruction that was achieved was relatively accurate, exhibiting a capability to distinguish the irregularities at the edges of the knob. Despite achieving an accuracy level of approximately 1 mm, it must be noted that these results were obtained under static conditions.
- In the event of a human entering the danger zone for inspection or replacement of the part, the robot immediately ceases its operations. A notable feature of this task provided by the ACROBA platform is the division of the zones into two quadrants, namely the right and left. This arrangement enhances the cycle time of the overall collaborative tasks

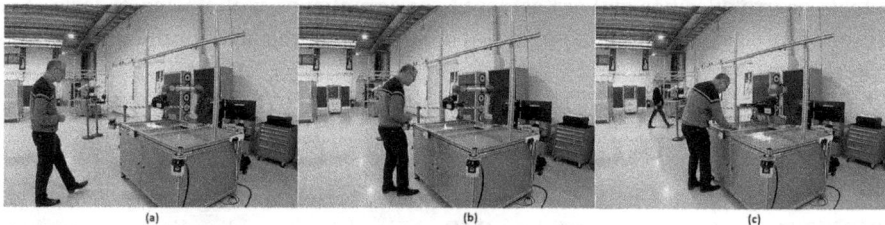

(a)   (b)   (c)

FIGURE 14.12 Human and robot collaborating to perform joint inspection task on SteriPack objects: (a) the robot scans objects for 3D reconstruction; (b) the human enters the danger zone, and the robot stops; (c) the human operator places another product for next inspection cycle.

while maintaining safety standards. This means that while a human is working in the right quadrant, the robot is permitted to continue its operation in the left quadrant and vice versa. This not only ensures a safe coexistence but also minimizes commissioning time.

14.7 SUMMARY

This chapter examines the implications and applications of modern digital tools in agile manufacturing. The discussion is prompted by the need to apply machine intelligence techniques to transform production systems into connected ecosystems, enabling improved product customization, optimization, and dissemination in the face of intrinsic and extrinsic variations. This leads to enhanced productivity and throughput. Industry 5.0 is introduced as a framework that supports human-centric development and exploitation of automation tools, promoting sustainability, flexibility, and reliability in connected ecosystems. The chapter further explores the use of autonomous mobile robots in agile manufacturing factories to enable flexible connections between the supply chain and production. Optimal scheduling and control are achieved through the application of a fleet management system. The chapter also elaborates on the relevant KPIs and metrics used to evaluate the performance of these tools and techniques in an agile ecosystem. The application domains of these KPIs and metrics are highlighted. The chapter also highlights the safety and ethical implications of utilizing modern digital tools in manufacturing industry. It presents the concepts governing these issues, including safety standards for maintaining dependability and ethics rules for ensuring trustworthiness.

Building on the theoretical discussion, this chapter analyzes the experimental outlook of the manufacturing industry that utilizes smart machines and methods. Two scenarios are considered. The first scenario involves the robotized assembly of agile parts, which entails the dynamic picking, transportation, and placement of candidate objects at designated positions to reduce cycle time and enhance the performance of the assembly line. The second scenario examines the automation of the inspection of manufactured parts, with human involvement. In this scenario, the manufactured parts undergo autonomous quality control to detect defects or irregularities and are assessed for compliance with design criteria. The presence of humans in this scenario improves uptime and implicitly reduces material wastage.

ACKNOWLEDGMENT

This work is supported by EU funded ACROBA project under grant agreement No 101017284. Authors are grateful to IMR, Ireland for providing research oriented environment and the opportunity to undertake this study with a view toward the existing manufacturing industry outlook.

REFERENCES

[1] Xu, X., Lu, Y., Vogel-Heuser, B. and Wang, L., 2021. Industry 4.0 and Industry 5.0—Inception, conception and perception. Journal of Manufacturing Systems, 61, pp. 530–535.

[2] https://www.3ds.com/manufacturing/connected-industry/industry-4-0-data-mass-customization-actionable-insights.

[3] European Commission, 2020b. Flash Eurobarometer 486: SMEs, start-ups, scale-ups and entrepreneurship.

[4] https://tulip.co/ebooks/agile-manufacturing/.

[5] Huang, S., Wang, B., Li, X., Zheng, P., Mourtzis, D. and Wang, L., 2022. Industry 5.0 and Society 5.0—Comparison, complementation and co-evolution. Journal of manufacturing systems, 64, pp. 424–428.

[6] https://ifr.org/downloads/press2018/2022_WR_extended_version.pdf

[7] Ogbemhe, J., Mpofu, K. and Tlale, N.S., 2017. Achieving sustainability in manufacturing using robotic methodologies. Procedia Manufacturing, 8, pp. 440–446.

[8] Dhirani, L.L., Mukhtiar, N., Chowdhry, B.S. and Newe, T., 2023. Ethical dilemmas and privacy issues in emerging technologies: A review. Sensors, 23(3), p.1151.

[9] Villani, V., Pini, F., Leali, F. and Secchi, C., 2018. Survey on human–robot collaboration in industrial settings: Safety, intuitive interfaces and applications. Mechatronics, 55, pp. 248–266.

[10] https://www.universal-robots.com/blog/is-automated-bin-picking-finally-real/#:~:text=The%20simple%20answer%20is%20that,a%20daunting%20task%20for%20robots

[11] Karagiannis, P., Kousi, N., Michalos, G., Dimoulas, K., Mparis, K., Dimosthenopoulos, D., Tokcalar, O¨., Guasch, T., Gerio, G.P. and Makris, S., 2022. Adaptive speed and separation monitoring based on switching of safety zones for effective human robot collaboration. Robotics and Computer-Integrated Manufacturing, 77, p. 102361

15 Role of Quantum Security in the Future of Smart Manufacturing

Danyal Maheshwari, Popenţiu-Vlădicescu Florin,
Lubna Luxmi Dhirani, Abi Waqas,
Bhawani Shankar Chowdhry, M. Mahmood Ali,
and Grigore Albeanu

15.1 INTRODUCTION

In the rapidly advancing world of technology, smart manufacturing is becoming an increasingly important field. With the integration of automation, data analysis, and connectivity, smart manufacturing has the potential to revolutionize the way that goods are produced and distributed [1]. However, as technology advances, so too do the potential risks and challenges that come with it. One major concern is the security of these systems, particularly in the realm of quantum computing [2]. This chapter will explore the role of quantum security in the future of smart manufacturing, and the implications that it has for the industry. Quantum computing has the potential to revolutionize the way that information is processed and stored. With its ability to perform complex calculations at an unprecedented speed and efficiency, quantum computing has the potential to greatly enhance the capabilities of smart manufacturing systems. However, as with any new technology, there are also potential security risks associated with quantum computing. One major concern is the potential for quantum computing to be used to crack encryption algorithms, potentially exposing sensitive information to unauthorized parties [1].

To address these concerns, researchers have begun to explore the use of quantum security techniques in smart manufacturing systems. These techniques include the use of quantum key distribution (QKD) to securely transmit information, as well as the use of quantum encryption to protect sensitive data. By implementing these techniques, smart manufacturing systems can be better protected against potential threats, ensuring the continued growth and success of the industry [2,3].

15.1.1 SMART MANUFACTURING

Smart Manufacturing is a term used to describe the integration of advanced technology, such as automation, data analysis, and connectivity, into the manufacturing process. It is a holistic approach that aims to improve efficiency,

DOI: 10.1201/9781003376620-20

productivity, and flexibility while reducing costs and environmental impact [3]. Smart manufacturing systems use sensors, internet of things (IoT) devices, and other technologies to collect real-time data on the production process. This data is then analyzed using advanced algorithms and machine learning techniques to optimize the production process and identify areas for improvement. Additionally, smart manufacturing systems can also be connected to other systems, such as supply chain management and logistics, to further optimize the production process. Smart manufacturing is also referred to as Industry 4.0, as it represents the fourth industrial revolution, following the mechanization of the 18th century, the mass production of the 19th century, and the automation of the 20th century. The goal of smart manufacturing is to create highly adaptable, efficient, and sustainable manufacturing systems that can respond quickly to changing market conditions and customer demands [4,5].

15.1.2 IMPORTANCE OF SECURITY IN SMART MANUFACTURING

Security is of paramount importance in smart manufacturing as it ensures the integrity, confidentiality, and availability of the manufacturing process. The integration of advanced technologies such as automation, data analysis, and connectivity in smart manufacturing systems increases the potential risk of cyber-attacks and unauthorized access to sensitive information [4].

The use of sensors, internet of things (IoT) devices, and other technologies in smart manufacturing systems, allows the collection of real-time data on the production process, making the systems vulnerable to cyber-attacks. An attacker with access to this data can manipulate the production process, cause downtime, and disrupt the entire supply chain. Furthermore, unauthorized access to intellectual property, trade secrets, and other sensitive information can lead to significant financial losses for the company. The smart manufacturing systems are also connected to other systems such as supply chain management and logistics, making them vulnerable to attacks that can spread beyond the manufacturing process. A cyber-attack on a smart manufacturing system can have a cascading effect on the entire supply chain, causing widespread disruption and significant financial losses. In addition, the integration of advanced technology in smart manufacturing systems increases the potential risks and challenges for the industry, highlighting the importance of security in smart manufacturing. Security measures such as access control, encryption, and monitoring are essential to protect the integrity, confidentiality, and availability of smart manufacturing systems [4–6].

15.2 OVERVIEW OF QUANTUM SECURITY AND ITS POTENTIAL ROLE IN SMART MANUFACTURING

Quantum security is a rapidly evolving field that aims to harness the power of quantum physics to create highly secure communication and encryption systems. With the advent of quantum computing, which can perform complex calculations at an unprecedented speed and efficiency, the need for quantum security techniques has become even more pressing. In smart manufacturing, quantum security techniques can be used to protect sensitive information such as intellectual property, trade secrets, and other confidential data. One of the key techniques used in quantum security is quantum key distribution (QKD), which allows for the secure transmission of information using quantum states. This technique is particularly useful in smart manufacturing systems, as it allows for the secure communication of data between different systems, such as production machines, supply chain management systems, and logistics systems [4,5].

Another key technique used in quantum security is quantum encryption, which uses the principles of quantum physics to encrypt data in a way that is virtually unbreakable. This technique can be used to protect sensitive information in smart manufacturing systems, ensuring that only authorized parties have access to it. In addition to these techniques, quantum security also includes quantum computing-based algorithms to identify and detect threats, as well as quantum-based access control mechanisms to ensure that only authorized individuals have access to smart manufacturing systems [5].

Quantum security has the potential to revolutionize the way that smart manufacturing systems are protected, providing a much higher level of security than traditional methods. As smart manufacturing systems continue to advance, the use of quantum security techniques will become increasingly important in ensuring the integrity and confidentiality of sensitive information.

15.2.1 Quantum Computing and Its Impact on Security

Quantum Computing is an innovative technology based on the principles of quantum mechanics. Unlike classical computers, quantum computers operate on quantum bits (qubits), which can store and process data in multiple states simultaneously, making them much faster and more efficient than classical computers. One of the most significant impacts of Quantum Computing is on security. It can quickly break many of the cryptographic algorithms currently in use, such as public-key encryption algorithms like RSA, making them vulnerable. This is because public-key encryption relies on the difficulty of factoring large numbers to keep messages secure. Quantum Computing has the potential to revolutionize several industries, including smart manufacturing, by enabling faster and more efficient processing of large amounts of data. It can be used to optimize manufacturing and supply chain processes, analyze data to improve product quality and predict equipment failures, and optimize energy usage in manufacturing facilities. Additionally, it can enhance cybersecurity by developing new encryption methods that are resilient to attacks from quantum computers [6] (Figure 15.1).

- **Manufacturing Optimization:** Quantum computing can be used to optimize manufacturing processes by simulating complex manufacturing scenarios and identifying the most efficient production methods.
- **Supply Chain Optimization:** Quantum computing can be used to optimize supply chain processes by predicting demand, optimizing inventory management, and reducing shipping times.

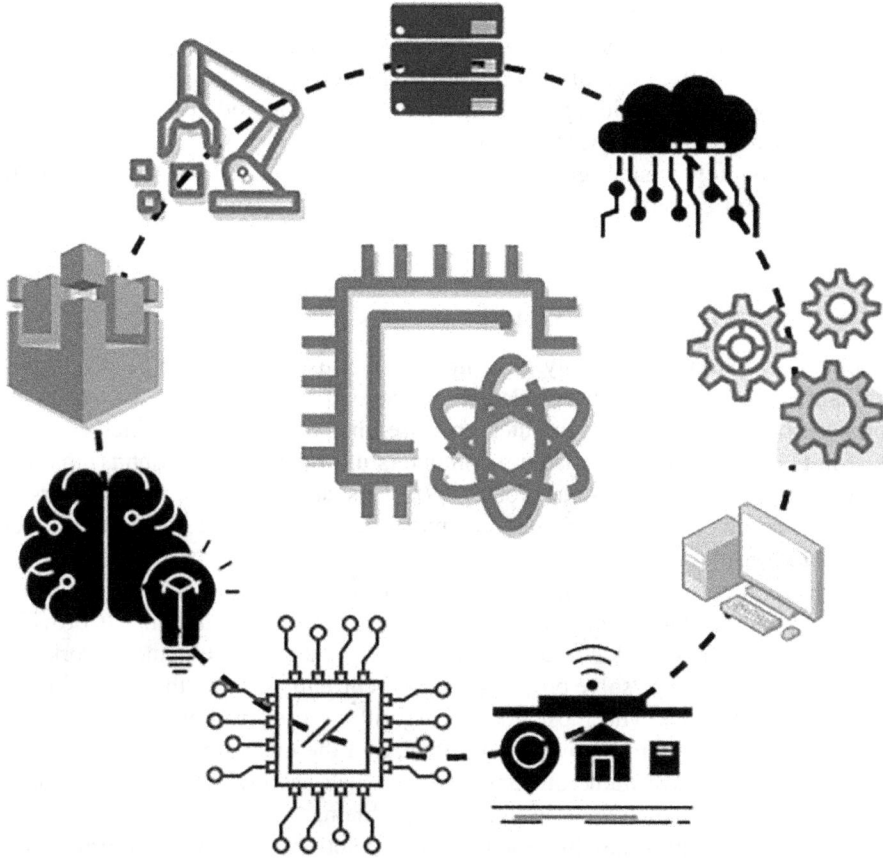

FIGURE 15.1 Quantum computing and its potential role in smart manufacturing.

- **Quality Control:** Quantum computing can be used to analyze large amounts of data from sensors and other sources to identify quality issues in real-time and improve product quality.
- **Predictive Maintenance**: Quantum computing can be used to analyze data from sensors and other sources to predict equipment failures before they occur, reducing downtime and improving efficiency.
- **Energy Optimization:** Quantum computing can be used to optimize energy usage in manufacturing facilities by identifying areas where energy consumption can be reduced.
- **Cybersecurity:** Quantum computing can be used to enhance cybersecurity in smart manufacturing by developing new encryption methods that are resistant to attacks from quantum computers.

Quantum Computing's impact on security includes the ability to perform quantum simulations, enabling the modeling of complex systems and real-time analysis to identify vulnerabilities and potential attacks. It also enables the development of new and more secure encryption algorithms, such as quantum key distribution protocols that securely transfer encryption keys between two parties. Additionally, Quantum Computing has the potential to revolutionize the field of security by providing new and more secure methods of encryption and enabling faster and more effective simulations of complex systems. However, it also presents new challenges that must be addressed, such as the need to upgrade current cryptographic algorithms and to develop new quantum-resistant algorithms [6–8].

15.3 QUANTUM COMPUTING

Quantum computing is a rapidly growing field that uses the principles of quantum mechanics to perform complex calculations beyond the capabilities of traditional computers. Unlike classical computers that use binary digits (bits), which can only exist in two states (0 or 1), quantum computers use quantum bits (qubits) that can exist in multiple states simultaneously [9]. This allows quantum computers to perform certain types of calculations much faster and more efficiently than traditional computers.

Quantum mechanics, which studies the behavior of matter and energy at the atomic and subatomic level, is the foundation of quantum computing. It provides the theoretical framework for understanding how quantum computers work and how they can be used to solve problems that are currently beyond the capabilities of traditional computers. The most notable algorithms used in quantum computing include Shor's algorithm, Grover's algorithm, and the quantum Fourier transform. These algorithms have numerous applications, such as code breaking, simulating quantum systems, and solving optimization problems.

Quantum computing has the potential to revolutionize the way information is processed and stored, opening up new opportunities for solving complex problems that are currently beyond the capabilities of conventional computers [10].

15.3.1 How Quantum Computing Differs from Classical Computing

Quantum computing is a rapidly growing field that utilizes the principles of quantum physics to perform calculations that surpass the capabilities of traditional computers. Unlike traditional binary digits (bits) that can only exist in two states, 0 or 1, quantum computers use quantum bits (qubits) that can exist in multiple states simultaneously [9]. This allows quantum computers to perform certain calculations much faster and more efficiently than traditional computers. For example, they can perform complex calculations, such as factoring large numbers and solving unstructured optimization problems, at a much faster rate than traditional computers [10].

Quantum mechanics, which is the branch of physics that studies the behavior of matter and energy at the atomic and subatomic level, provides the theoretical framework for understanding how quantum computers work and how they can be used to solve problems that are currently beyond the capabilities of traditional computers. Some of the popular algorithms that run on quantum computers include Shor's algorithm, Grover's algorithm, and the quantum Fourier transform, which can be applied to a wide range of applications such as code breaking, simulating quantum systems, and optimization problems.

Quantum computing has the potential to revolutionize the way information is processed and stored, providing new opportunities for solving complex problems that are currently beyond the capabilities of conventional computers [10].

15.3.2 POTENTIAL BENEFITS OF QUANTUM COMPUTING FOR SECURITY IN SMART MANUFACTURING

Quantum computing has the potential to revolutionize the field of security in smart manufacturing. The following are some of the benefits of quantum computing in this regard (Figure 15.2):

- **Improved Encryption:** Quantum computing can provide more secure encryption algorithms than classical computers. These algorithms are based on mathematical problems that are difficult to solve, even for quantum computers. This means that the data being transmitted or stored can be protected from unauthorized access.
- **Faster Processing:** Quantum computing can process vast amounts of data at much faster speeds than classical computers. This can be beneficial in security-critical applications where fast processing times are necessary, such as in real-time threat detection.
- **Advanced Simulation:** Quantum computing can simulate complex systems that classical computers cannot. This can be used to test and validate security systems, and to identify weaknesses that can be addressed before they are exploited by attackers.

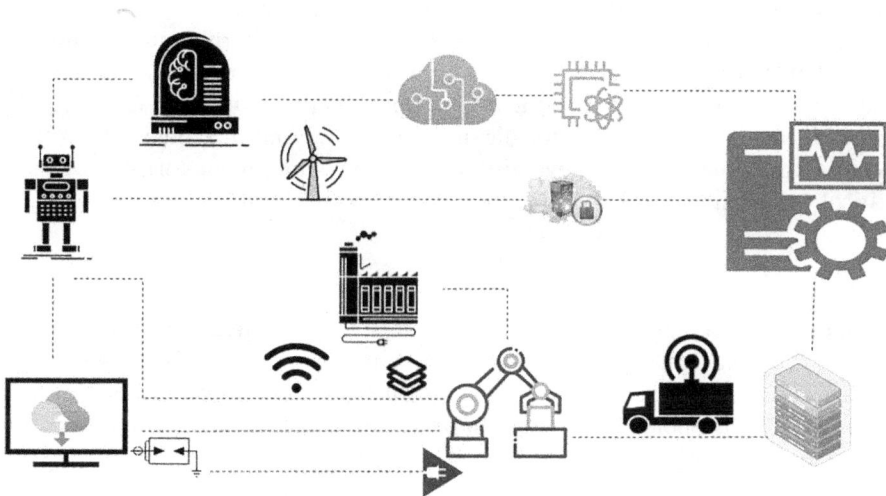

FIGURE 15.2 Quantum cryptography and its role in smart manufacturing.

- **Improved Analytics:** Quantum computing can process large amounts of data quickly and accurately, making it easier to identify patterns and trends. This can be used to analyze security data, such as network traffic, to identify potential threats and respond to them in real time.
- **Increased Complexity:** Quantum computing can handle complex algorithms that classical computers cannot. This can be used to design more secure systems, such as secure key distribution systems and secure communication networks [11–13].

15.4 QUANTUM CRYPTOGRAPHY AND ITS ROLE IN SMART MANUFACTURING

Quantum cryptography is a technique that uses quantum mechanical properties such as superposition and entanglement to secure communications. Unlike traditional cryptography, which uses mathematical algorithms that can be broken with enough computational power, quantum cryptography is based on the principles of quantum mechanics and is much more secure [12,13].

In smart manufacturing, where sensitive data is transmitted between different devices and systems, quantum cryptography plays a critical role in ensuring data security. Quantum cryptography can be used in various areas like manufacturing, IoT, transport, healthcare, cloud computing, robotics, and monitoring systems, to provide secure communication between different devices and prevent unauthorized access to sensitive information. For example, in manufacturing, quantum cryptography can secure communication between different parts of a production line, preventing the leakage or compromise of sensitive data. In healthcare, it can ensure that patient data remains confidential during medical procedures. In robotics, it can secure communication between different robots and systems to prevent attacks on the communication network and protect sensitive data about the robotics system [14,15].

Quantum cryptography can play an essential role in smart manufacturing and other fields where sensitive data is transmitted between different devices and systems. In smart manufacturing specifically, quantum cryptography provides a secure network for devices, sensors, and machines to collect, process, and control data, preventing cyber-attacks and unauthorized access to critical information [15].

15.4.1 QUANTUM CRYPTOGRAPHY

Quantum cryptography refers to a field of study that deals with the use of quantum mechanical systems to secure communication. It is a method of transmitting secure messages by encoding them in the properties of quantum particles such as photons. The concept of quantum cryptography is based on the principle of quantum mechanics, which states that the mere act of observing a quantum system changes its state. This means that any attempt to intercept the message will cause the quantum state of the particles to change, thereby alerting the sender and receiver of the message that their communication has been compromised [16].

Quantum cryptography is one of the most secure methods of communication as it is virtually impossible to intercept the message without being detected. This is because of the laws of quantum mechanics which make it impossible to observe a quantum system without causing a change in its state. One of the most well-known applications of quantum cryptography is quantum key distribution (QKD), which uses quantum mechanics to securely distribute encryption keys. This method is based on the principle that the mere act of observing a quantum system changes its state. In QKD, a random key is generated and transmitted between two parties through a quantum channel. The key is then used to encrypt the message [17].

15.4.2 How Quantum Cryptography Differs from Classical Cryptography

Quantum cryptography and classical cryptography are both techniques used to secure communication and encryption systems; however, they differ in the way that they provide security. Classical cryptography is based on mathematical algorithms and mathematical problems that are hard to solve [17,18].

The most common classical cryptography techniques include symmetric-key encryption, asymmetric-key encryption, and hash functions. These techniques rely on the difficulty of solving mathematical problems, such as factoring large numbers, to provide security. On the other hand, quantum cryptography is based on the principles of quantum physics. It uses the properties of quantum states, such as superposition and entanglement, to provide security. The most common quantum cryptography techniques include quantum key distribution (QKD) and quantum encryption. These techniques rely on the laws of quantum physics, such as the Heisenberg uncertainty principle, to provide security [18].

One of the key differences between quantum cryptography and classical cryptography is that quantum cryptography is unbreakable, even by a quantum computer. This is because any attempt to measure the key in QKD will introduce errors, and the parties will be able to detect these errors and discard the key. In contrast, classical cryptography can be broken by a powerful enough computer using a brute force attack. Another key difference is that classical cryptography relies on mathematical problems that can be solved by a computer, whereas quantum cryptography relies on physical properties that cannot be simulated by a computer. This makes quantum cryptography more secure and robust against hacking and other forms of attacks [19–21]. In conclusion, quantum cryptography and classical cryptography are both techniques used to secure communication and encryption systems, but they differ in the way that they provide security. Quantum cryptography is more secure and robust against attacks, as it is based on the principles of quantum physics, which are unbreakable by a computer [22].

15.4.3 POTENTIAL APPLICATIONS OF QUANTUM CRYPTOGRAPHY IN SMART MANUFACTURING

Quantum cryptography has the potential to revolutionize the field of security in smart manufacturing in several ways. The unique properties of quantum states, such as superposition and entanglement, provide a level of security that is virtually unbreakable, even by a quantum computer. This makes quantum cryptography an ideal solution for secure communication in smart manufacturing systems. One of the most significant applications of quantum cryptography in smart manufacturing is in the protection of intellectual property and trade secrets. Smart manufacturing systems generate a large amount of data, including designs, production data, and sensor data, which are critical to the manufacturing process. Quantum cryptography can be used to encrypt this data, ensuring that it remains confidential and that only authorized parties have access to it [20].

Another potential application of quantum cryptography in smart manufacturing is in the protection of the communication between different systems, such as production machines, supply chain management systems, and logistics systems. Quantum key distribution (QKD) can be used to securely transmit information between these systems, ensuring that the data remains confidential and that it is not intercepted by unauthorized parties. In addition, quantum cryptography can also be used to provide secure access control mechanisms in smart manufacturing systems. Quantum-based access control mechanisms can be used to ensure that only authorized individuals have access to the systems, and to detect malicious actors in the smart manufacturing systems. Furthermore, quantum cryptography can be used to provide secure communication in the IoT systems that are being integrated into smart manufacturing systems. IoT systems often rely on wireless communication, which is vulnerable to interception and hacking. Quantum cryptography can be used to secure communication between IoT devices, ensuring that the data remains confidential and that it is not intercepted by unauthorized parties [22,23].

In addition, quantum cryptography has the potential to revolutionize the field of security in smart manufacturing by providing highly secure communication and encryption systems, new algorithms for identifying and detecting threats, and optimization of the smart manufacturing process [23].

15.5 QUANTUM KEY DISTRIBUTION (QKD) AND ITS IMPORTANCE IN SMART MANUFACTURING

Quantum Key Distribution (QKD) is a technique that harnesses the principles of quantum physics to secure communication and encryption systems. It is considered one of the most secure methods of communication and encryption as it is virtually unbreakable by even the most advanced quantum computers. QKD enables the secure transmission of information by encoding the key in quantum states of light. QKD is critical technology for securing data transmission in various fields, including manufacturing, IoT, transport, healthcare, hospitals, cloud computing, robotics, financial institutes, national databases, energy, and monitoring systems. It allows two parties to establish a shared secret key over an insecure channel, such as the internet, without any risk of interception or eavesdropping. Any attempt to intercept the key will inevitably disturb the quantum states, alerting the parties to the presence of an eavesdropper [16,17].

The importance of QKD in smart manufacturing and other areas lies in its ability to provide unbreakable security for sensitive data. In manufacturing, QKD can be used to protect intellectual property, trade secrets, and other confidential information. In IoT, QKD can secure communications between devices and prevent unauthorized access to sensitive data. In transport, QKD can be used to protect data transmitted between vehicles, such as location and status information. In healthcare and hospitals, QKD can help protect patient records and other sensitive information. In cloud computing, QKD can be used to secure data transmitted between different parts of a distributed system. In robotics, QKD can be used to secure communication between robots and prevent unauthorized access or tampering with sensitive data. In financial institutes, QKD can help secure transactions and protect against fraud. In national databases, QKD can be used to protect sensitive information about citizens and prevent unauthorized access or tampering. In energy and monitoring systems, QKD can be used to secure communication between sensors and prevent unauthorized access to sensitive data [16,17,20,21].

In QKD, a secret key is generated and shared between two parties, Alice and Bob, through the use of quantum states. The key can then be used to encrypt and decrypt information, ensuring only authorized parties have access to it. The key is generated by transmitting a sequence of single photons through an optical channel and measuring them at the receiver's end. The measurement of the photons randomizes the key, and the parties can use it for encryption and decryption of information. The security of QKD is based on the principles of quantum mechanics, specifically on the Heisenberg uncertainty principle, which states that the more precisely the position of a particle is known, the less precisely its momentum can be known, and vice versa (Figure 15.3).

This principle ensures that any attempt to measure the key will introduce errors, which the parties can detect and discard the key. QKD is particularly important in smart manufacturing systems, as it enables the secure communication of data between different systems, such as production machines, supply chain management systems, and logistics systems. In smart manufacturing systems, sensitive data such as designs, production data, and sensor data needs protection to avoid intellectual property theft, unauthorized access, and cyber-attacks. QKD provides a secure communication channel between different systems, ensuring that the data remains confidential and is not intercepted by unauthorized parties. Furthermore, QKD can also provide secure access control mechanisms in smart manufacturing systems, ensuring that only authorized individuals have access to the systems and detecting malicious actors [24].

FIGURE 15.3 Quantum key distribution and its importance in smart manufacturing.

15.5.1 QUANTUM KEY DISTRIBUTION

Quantum key distribution (QKD) is a technique used to securely transmit information through an optical channel, by using the principles of quantum mechanics. It is one of the most secure methods of communication and encryption, and it is particularly useful in securing data and communications in a wide range of applications, such as financial transactions, military communications, and smart manufacturing systems [25].

In QKD, a secret key is generated and shared between two parties, known as Alice and Bob, using quantum states. This key can then be used to encrypt and decrypt information, ensuring that only authorized parties have access to it. The key is generated by transmitting a sequence of single photons through an optical channel and measuring them at the receiver's end. The measurement of the photons randomizes the key, and the parties can use it for encryption and decryption of information. The security of QKD is based on the principles of quantum mechanics, specifically on the Heisenberg uncertainty principle, which states that the more precisely the position of a particle is known, the less precisely its momentum can be known, and vice versa. This principle ensures that any attempt to measure the key will introduce errors, and the parties will be able to detect these errors and discard the key [15,16]. One of the main advantages of QKD is that it provides a secure means of communication, even in the face of a powerful quantum computer. This is because any attempt to measure the key will introduce errors, and the parties will be able to detect these errors and discard the key. This makes it an ideal solution for protecting sensitive information, such as intellectual property and trade secrets, in smart manufacturing systems. The QKD is a secure method of transmitting information through an optical channel, by using the principles of quantum mechanics. It is considered to be one of the most secure methods of communication and encryption, and it is particularly useful in a wide range of applications, such as smart manufacturing systems [25].

15.5.2 Working of Quantum Key Distribution

Quantum key distribution (QKD) is a technique used to secure communication and encryption systems by harnessing the principles of quantum physics. It allows for the secure transmission of information using quantum states. The process of QKD can be broken down into several steps:

- **Key Generation:** The first step in QKD is to generate a secret key that will be used to encrypt and decrypt information. The key is generated by transmitting a sequence of single photons through an optical channel and measuring them at the receiver's end. The measurement of the photons randomizes the key, and the parties can use it for encryption and decryption of information.
- **Key Distribution:** Once the key is generated, it is distributed between the two parties, known as Alice and Bob. This is done by transmitting the key over an optical channel, such as a fiber optic cable. The key is encoded on the photons, and the parties can measure the photons to extract the key.
- **Error Detection:** During the transmission of the key, errors may occur due to noise or other factors. To detect these errors, Alice and Bob use a technique called error correction, which allows them to detect and discard any errors in the key.
- **Key Authentication:** Once the key is distributed and any errors have been detected and corrected, the parties can authenticate the key. This is done by comparing the key to a pre-agreed value, called a check value, which is generated by both parties before the key is distributed.
- **Key Encryption and Decryption:** Once the key has been authenticated, it can be used to encrypt and decrypt information. The key is used to encrypt the information at the sender's end, and to decrypt the information at the receiver's end.

The QKD is a secure technique that allows for the transmission of information through an optical channel, by using the principles of quantum mechanics.

It is considered to be one of the most secure methods of communication and encryption, as it is virtually unbreakable by even the most advanced quantum computers [16].

15.5.3 POTENTIAL BENEFITS OF QUANTUM KEY DISTRIBUTION FOR SECURITY IN SMART MANUFACTURING

Quantum Key Distribution (QKD) offers a new level of security for smart manufacturing. It provides a secure communication channel for transmitting data, even in the presence of malicious actors [12,13,15]. The potential benefits of QKD for smart manufacturing include:

- **Unbreakable security:** QKD relies on the laws of quantum physics to secure the transmission of data. This makes it impossible for attackers to eavesdrop or manipulate the data, ensuring that sensitive information remains confidential.
- **Tamper-proof authentication:** QKD enables secure authentication of users, devices, and transactions. This helps to prevent tampering and counterfeiting of products and reduces the risk of theft or fraud.
- **Improved efficiency:** QKD can be integrated into existing networks, eliminating the need for additional hardware and software. This results in increased efficiency and reduces the time and cost of data transmission.
- **Increased resilience:** QKD is highly resilient to cyber-attacks, such as hacking and denial-of-service (DoS) attacks. This helps to ensure that the network remains operational even in the event of an attack, preventing the loss of valuable data.

15.6 CHALLENGES AND LIMITATIONS OF QUANTUM SECURITY IN SMART MANUFACTURING

Quantum security has the potential to revolutionize the field of security in smart manufacturing; however, it also faces several challenges and limitations. One of the main challenges is the lack of infrastructure and standardization for quantum communication and encryption systems. Currently, there are few commercial quantum communication systems available, and there is a lack of standardization among different systems, which can make it difficult to integrate quantum security into smart manufacturing systems. Another challenge is the cost and complexity of implementing quantum security systems. Quantum communication and encryption systems are still in the early stages of development and can be costly to implement and maintain. The complexity of these systems can also make it difficult for manufacturers to fully understand and utilize their capabilities. Furthermore, the lack of scalability of quantum security systems. Quantum communication and encryption systems are currently limited in their range and capacity, making it difficult to secure large-scale manufacturing systems [26,27].

There are also limitations to the level of security that quantum security systems can provide. Quantum security systems are still vulnerable to certain types of attacks, such as side-channel attacks, which exploit physical character-istics of the system, such as power consumption or electromagnetic radiation, to extract secret keys. In addition, the integration of quantum security systems with existing classical security systems can also be a challenge. This is because, quantum security systems are based on a different set of principles and technologies than classical security systems, and this can make it difficult to integrate them seamlessly. While quantum security has the potential to revolutionize the field of security in smart manufacturing, there are still many challenges and limitations that need to be overcome before it can be fully realized [28].

15.6.1 CHALLENGES OF QUANTUM SECURITY IN SMART MANUFACTURING

- **Integration with existing systems:** Integrating quantum security into existing smart manufacturing systems is a complex task that requires significant technical skills and resources.
- **Implementation cost:** Implementing quantum security solutions in smart manufacturing can be expensive due to the cost of hardware, software, and training.
- **Technical complexity:** Quantum security solutions can be difficult to understand and implement, even for experienced professionals, due to the technical complexity of quantum mechanics.
- **Scalability:** Scalability is a major challenge in quantum security, as the number of nodes in a quantum network increases, the complexity of managing and maintaining the network also increases.
- **Interoperability:** Interoperability is another challenge in quantum secu-rity, as different quantum security solutions may not be compatible with each other, making it difficult to integrate different quantum security systems in smart manufacturing.
- **Limited availability of quantum technology:** Currently, quantum tech-nology is not widely available, and its implementation is limited to specialized applications and research institutions.
- **Unpredictable results:** The unpredictable nature of quantum mechanics can result in unpredictable results in quantum security systems, which can lead to security vulnerabilities.
- **Vulnerability to quantum attacks:** Despite being highly secure, quantum security systems are vulnerable to quantum attacks, which can exploit weaknesses in the quantum algorithms used in quantum security systems.
- **Inflexibility:** Quantum security systems are inflexible and cannot be easily modified once they are implemented, making it difficult to adapt to changing security requirements in smart manufacturing [29].

15.6.2   TECHNICAL CHALLENGES OF IMPLEMENTING QUANTUM SECURITY IN SMART MANUFACTURING

Quantum security has the potential to revolutionize the field of security in smart manufacturing; however, it also faces several technical challenges that need to be overcome before it can be fully implemented [30–32]. One of the main challenges is the lack of infrastructure and standardization for quantum communication and encryption systems. Currently, there are few commercial quantum communication systems available, one of the reasons for this is lack of standardization in enabling technologies, that leads to complexity in quantum security integrations in the smart manufacturing environment. Another issue is the cost and complexity of implementing quantum security systems. Quantum communication and encryption systems are still in the early stages of development and can be costly to implement and maintain. The complexity of these systems can also make it difficult for manufacturers to fully understand and utilize their capabilities. Lack of scalability of quantum security systems is another limitation, quantum communication and encryption systems are currently constrained in their range and capacity, making it difficult to secure large-scale manufacturing systems [33,34].

Furthermore, there are also limitations to the level of security that quantum security systems can provide. Quantum security systems are still vulnerable to certain types of attacks, such as side-channel attacks, which exploit physical characteristics of the system, such as power consumption or electromagnetic radiation, to extract secret keys [35]. In addition, the integration of quantum security systems with existing classical security systems can also be a challenge. This is because, quantum security systems are based on a different set of principles and technologies than classical security systems, and this can make it difficult to integrate them seamlessly. Overall, while quantum security can transform the security aspects in smart manufacturing, there are still many technical challenges that need to be mitigated before it can be implemented [35–37].

15.6.3   LIMITATIONS OF CURRENT QUANTUM TECHNOLOGY

Quantum technology can enable the smart manufacturing environment by providing highly secure communication and encryption systems, new algorithms for identifying and detecting threats, and optimization of the manufacturing process [38,39]. However, current quantum technology also has several limitations (as discussed in section 5.1), that need to be overcome before it can be deployed in smart manufacturing systems. The first is limitations associated with scalability, making it a poor choice for securing large-scale manufacturing systems. The current technology is based on the use of single photons, which can be affected by environmental noise and other factors, leading to errors in the transmission of information. Second, maintenance costs would be high due to limited skill set and early development phases of encryption and communication systems [39–41]. Third, the technologies are vulenrable toward certain types of cyber-attacks, leading to potential exploitation of the physical characteristics of the system. Furthermore, integrating quantum technology with existing/conventional technology is challenging as well, as it is based on a set of 15 different principles that make it complex to integrate seamlessly [35,42,43]. While quantum may have the potential to revolutionize the smart manufacturing ecosystem, there are still many limitations that need to be overcome before it can be fully realized [38,44].

15.6.4 POTENTIAL FUTURE DEVELOPMENTS IN QUANTUM SECURITY

Quantum security has the potential to revolutionize the field of security in a wide range of applications, including smart manufacturing systems. While current quantum security technology still faces several limitations, there are several potential future developments that could help overcome these limitations and bring quantum security to its full potential. One potential future development is the use of quantum repeaters. Quantum repeaters are devices that can amplify and extend the range of quantum communication systems. This would make it possible to secure large-scale manufacturing systems by allowing for the secure transmission of information over greater distances [43,44].

The potential future development would be using quantum error correction codes, enabling error detection and correction during transmission of information in quantum communication systems. This would make quantum security systems more resilient to environmental noise and other factors that could lead to errors while transmitting information. Furthermore, the development of quantum-safe cryptographic algorithms can also be a potential future development. These algorithms can be used to encrypt data even in the face of a powerful quantum computer, which would provide an added layer of security in smart manufacturing systems [45]. While current quantum security technology still faces several limitations, there are several potential future developments that could help overcome these limitations and bring quantum security to its full potential [46,47].

15.7 QUANTUM SMART INDUSTRY 1.0

Quantum Smart Industry 1.0 refers to the application of quantum technologies in the industrial sector to enhance the efficiency and productivity of various processes. This integration of quantum computing, quantum communication, and quantum sensing technologies into the industry is aimed at solving complex industrial problems and optimizing operations. The integration of quantum technology in the industry is expected to bring significant benefits in terms of improved performance, reduced costs, and increased competitiveness [48].

Quantum computing is expected to play a critical role in the development of Quantum Smart Industry 1.0. The processing power of quantum computers is significantly higher than that of classical computers, making them ideal for solving complex problems in various industrial sectors. For instance, quantum computers can be used to optimize supply chain management, reduce energy consumption, and improve product design. In the pharmaceutical industry, quantum computers can be used to speed up the discovery of new drugs, as well as predict their efficacy and toxicity [49].

Quantum communication is another critical component of Quantum Smart Industry 1.0. This technology provides secure communication channels for industrial processes and transactions. This is essential for the protection of sensitive industrial data and intellectual property, which is critical for the success of businesses in the 21st century.

Quantum sensing technologies are also expected to play a significant role in Quantum Smart Industry 1.0. These technologies enable real-time measurement of physical parameters, such as temperature, pressure, and magnetic fields, which is essential for various industrial processes. For example, quantum sensors can be used to monitor the quality of industrial products, as well as to detect faults in production processes [49,50].

In conclusion, the integration of quantum technologies in the industrial sector is expected to bring significant benefits in terms of improved efficiency and productivity. The development of Quantum Smart Industry 1.0 is expected to be a key driver of innovation and competitiveness in the global economy, as well as a major contributor to the growth of the technology sector [51].

15.8   CONCLUSION

15.8.1   Summary of the Potential Role of Quantum Security in Smart Manufacturing

Quantum security is a rapidly emerging technology that has the potential to revolutionize the field of security in smart manufacturing systems. The unique properties of quantum states, such as superposition and entanglement, provide a level of security that is virtually unbreakable, even by a powerful quantum computer. This makes quantum security an ideal solution for secure communication and encryption in smart manufacturing systems. Quantum key distribution (QKD) is a technique that is particularly useful in smart manufacturing systems. It allows for the secure communication of data between different systems, such as production machines, supply chain management systems, and logistics systems. The security of QKD is based on the principles of quantum mechanics, specifically on the Heisenberg uncertainty principle, which ensures that any attempt to measure the key will introduce errors, and the parties will be able to detect these errors and discard the key. Quantum cryptography also has the potential to provide secure access control mechanisms in smart manufacturing systems. Quantum-based access control mechanisms can be used to ensure that only authorized individuals have access to the systems, and to detect malicious actors. However, current quantum security technology still faces several limitations, such as lack of scalability, cost and complexity, vulnerability to certain types of attacks, and integration with existing classical technology. But future developments such as the use of quantum repeaters, quantum error correction codes, quantum key distribution protocols, and quantum-safe cryptographic algorithms can overcome these limitations and bring quantum security to its full potential.

In short, quantum security has the potential to revolutionize the field of security in smart manufacturing by providing highly secure communication and encryption systems, new algorithms for identifying and detecting threats, and optimization of the smart manufacturing process.

15.8.2 Outlook for the Use of Quantum Security in Smart Manufacturing

The outlook for the use of quantum security in smart manufacturing is extremely promising, as it has the potential to revolutionize the field of security in smart manufacturing systems. The unique properties of quantum states, such as superposition and entanglement, provide a level of security that is virtually unbreakable, even by a powerful quantum computer. This makes quantum security an ideal solution for secure communication and encryption in smart manufacturing systems. In the future, quantum key distribution (QKD) is expected to play a significant role in securing data and communication in smart manufacturing systems. QKD can be used to encrypt data, ensuring that it remains confidential and that only authorized parties have access to it. The scalability of QKD is also expected to improve in the future, allowing for the secure communication of data between different systems, such as production machines, supply chain management systems, and logistics systems. Quantum cryptography is also expected to be an essential part of smart manufacturing security, providing secure access control mechanisms and secure communication in IoT systems that are being integrated into smart manufacturing systems.

In addition, the integration of quantum security with artificial intelligence (AI) and machine learning (ML) is also expected to be an important area of development in the future. This can be used to improve the efficiency and security of smart manufacturing systems by providing real-time monitoring and analysis of data resulting in identifying and detecting threats in real-time. As such, the outlook for the use of quantum security in smart manufacturing is extremely promising. With the continued development and advancement of quantum technology, it is expected that quantum security will become an essential part of smart manufacturing systems, providing highly secure communication and encryption systems, new algorithms for identifying and detecting threats, and optimization of the smart manufacturing process.

REFERENCES

[1] Gisin, N., Ribordy, G., Tittel, W., Zbinden, H. (2002). Quantum cryptography. Reviews of Modern Physics, 74(1), 145.
[2] Ladd, T. D., Jelezko, F., Laflamme, R., Nakamura, Y., Monroe, C., O'Brien, J. L. (2010). Quantum computers. Nature, 464(7285), 45–53.
[3] Dijk, M., Pinto, J. (2016). Quantum key distribution in smart grid systems. IEEE Transactions on Smart Grid, 7(5), 2132–2139.
[4] Dhirani, L. L., Eddie, A., Thomas, N. (2021). Industrial IoT, cyber threats, and standards landscape: Evaluation and roadmap. Sensors, 21(11), 3901.
[5] Sajeed, S., Jain, A. (2017). Quantum cryptography in smart grid: A review. IEEE Access, 5, 20816–20831.

[6] Mower, J., Riggio, R., Verma, V., Sadeghi, A., Ruan, H., Tippenhauer, N. (2015). On the security of quantum key distribution in the presence of a continuous-variable eavesdropper. IEEE Journal of Selected Topics in Quantum Electronics, 21(3), 1–10.

[7] Wahlster, W., Rieder, B. (2015). Smart manufacturing: Industry 4.0—the future of productivity and growth in manufacturing industries. Journal of Manufacturing Systems, 34(1), 1–3.

[8] Mathew, J., Ramachandran, M. (2018). Smart manufacturing: A review of research and developments. Journal of Manufacturing Systems, 47, 1–14.

[9] Chen, Y., Kao, H. (2015). A review of smart manufacturing systems and applications. Manufacturing Letters, 3, 1–7.

[10] Fagiano, L., Boffi, A. (2019). Smart manufacturing: A review of the state-of-the-art and future perspectives. Journal of Manufacturing Systems, 52, 1–20.

[11] Zhang, X., Liu, X., Guo, Q. (2019). Smart manufacturing: A review of the research and development trend. IEEE Access, 7, 134866–134879.

[12] Riggio, R., Verma, V., Sadeghi, A., Tippenhauer, N. (2016). Security and privacy challenges in smart manufacturing systems. IEEE Communications Magazine, 54(9), 24–30.

[13] Chen, Y., Kao, H. (2015). A review of smart manufacturing systems and applications. Manufacturing Letters, 3, 1–7.

[14] Nielsen, M. A., Chuang, I. L. (2010). Quantum Computation and Quantum Information (10th ed.). Cambridge University Press.

[15] Lidar, D. A., Brun, T. A. (Eds.). (2013). Quantum Error Correction. Cambridge University Press.

[16] Arute, F., Arya, K., Babbush, R., Bacon, D., Bardin, J. C., Barends, R., ... Mandr`a, S. (2019). Quantum supremacy using a programmable superconducting processor. Nature, 574(7779), 505–510.

[17] Bergholm, V., Djordjevic, I. B., Fitzi, M. (2018). Quantum computing and quantum cryptography. Nature Communications, 9(1), 4366.

[18] Deutsch, D. (1985). Quantum theory, the Church-Turing principle, and the universal quantum computer. Proceedings of the Royal Society of London A, 400(1818), 97–117.

[19] Shor, P. W. (1994). Algorithms for quantum computation: Discrete logarithms and factoring. In Proceedings of the 35th Annual Symposium on Foundations of Computer Science (pp. 124–134). IEEE Computer Society Press.

[20] Grover, L. K. (1997). Quantum mechanics helps in searching for a needle in a haystack. Physical Review Letters, 79(2), 325–328.

[21] Coppersmith, D. (1994). An approximate Fourier transform useful in quantum factoring. IBM Research Report RC 19642.

[22] Riggio, R., Verma, V., Sadeghi, A., Tippenhauer, N. (2016). Security and privacy challenges in smart manufacturing systems. IEEE Communications Magazine, 54(9), 24–30.

[23] Duan, L. M., Guo, G. C. (1997). Quantum entanglement and quantum cryptography. In Progress in Optics (pp. 378–456). Elsevier.

[24] Maheshwari, D., Garcia-Zapirain, B., Sierra-Soso, D. (2020). Machine learning applied to diabetes dataset using Quantum versus Classical computation. IEEE Int. Symp. Signal Process. Inf. Technol. ISSPIT 2020, 2020.

[25] Maheshwari, D., Sierra-Sosa, D., Garcia-Zapirain, B. (2022). Variational quantum classifier for binary classification: Real vs synthetic dataset. IEEE Access, 10, 3705–3715.

[26] Maheshwari, D., Garcia-Zapirain, B., Sierra-Sosa, D. (2022). Quantum machine learning applications in the biomedical domain: A systematic review. in IEEE Access, 10, 80463–80484. doi: 10.1109/ACCESS.2022.3195044.

[27] Ullah, U., Maheshwari, D., Gloyna, H. H., Garcia-Zapirain, B. (2022). Severity classification of COVID-19 patients data using quantum machine learning approaches. 2022 International Conference on Electrical, Computer, Communications and Mechatronics Engineering (ICECCME), Maldives, Maldives, pp. 1–6, doi: 10.1109/ICECCME55909.2022.9987991

[28] D. Maheshwari, U. Ullah, Pablo A. O. M., A. Garc´ıa-Olea Jurado, I. Diez Gonzalez, J. M. Ormaetxe Merodio, B. Garcia-Zapirain. (2023). Quantum machine learning applied to electronic healthcare records for ischemic heart disease classification, Article number: 13:06

[29] Ekert, A. K. (1991). Quantum cryptography based on Bell's theorem. Physical Review Letters, 67(6), 661–663.

[30] Pirandola, S., Banchi, L., Biswas, A., Braunstein, S. L. (2015). Advances in quantum cryptography. Nature Photonics, 9(6), 641–652.

[31] Zeng, H. (2017). A review on quantum cryptography and its applications in smart manufacturing. Journal of Ambient Intelligence and Humanized Computing, 8(2), 149–157.

[32] Scarani, V., Acin, A., Ribordy, G., Gisin, N. (2009). The security of practical quantum key distribution. Reviews of Modern Physics, 81(3), 1301.

[33] "Quantum Cryptography." Stanford Encyclopedia of Philosophy, Stanford University, plato.stanford.edu/entries/quantum-cryptography/

[34] "Quantum Cryptography." Wikipedia, Wikimedia Foundation, 18 Oct. 2021, en.wikipedia.org/wiki/Quantumcryptography

[35] "Quantum Key Distribution (QKD)." NIST, National Institute of Standards and Technology, csrc.nist.gov/Projects/Quantum-Based-Cryptography/Quantum-csrc.nist.gov/Projects/Quantum-Based-Cryptography/Quantum-Key-Distribution

[36] "Quantum Cryptography." Physicsworld.com, Institute of Physics, physicsworld.com/a/what-is-quantum-cryptography/

[37] Wang, X. B. (2017). Quantum cryptography. In Quantum Information and Quantum Computing (pp. 1–28). Springer, Singapore.

[38] Quantum Computing and Information Security. (2021, May 17). https://www.rsa.com/en-us/insights/research/quantum-computing-and-information-security.

[39] Quantum Computing and its Implications for Cybersecurity. (2021, May 21). https://www.weforum.org/agenda/2021/05/quantum-computing-and-its-implications-for-cybersecurity/.

[40] Quantum Computing and Cybersecurity: Benefits and Risks https://www.ibm.com/blogs/research/2020/11/quantum-computing-and-cybersecurity-benefits-and-risks. (2020, November 4).

[41] Quantum Cryptography: A Brief Overview, NASA: https://www.nasa.gov/centers/kennedy/technology/technology-features/Quantum-Cryptography.html

[42] The Benefits of Quantum Key Distribution, RSK Labs: https://rsk.co/blog/the-benefits-of-quantum-key-distribution/

[43] Kim, Y., Kim, J. (2017). A survey on quantum key distribution: principles and security. arXiv preprint arXiv:1703.03104.

[44] Konuma, M., Aonuma, Y., Fumoto, K. (2017). Smart manufacturing: challenges and trends. Journal of Ambient Intelligence and Humanized Computing, 8(2), 161–169.

[45] Nie, Y., Lu, R. (2019). Research on quantum cryptography technology and its application in smart manufacturing. Journal of Ambient Intelligence and Humanized Computing, 10(3), 417–423.

[46] Peters, N. A., Dufour, G. D., Kwiat, P. G. (2004). Demonstration of high-efficiency quantum teleportation using pulsed entangled photons. Nature, 429(6994), 161–164.

[47] Lo, H.-K., Curty, M. (2014). Secure quantum key distribution. Nature Photonics, 8(8), 595–604.

[48] Jain, J. (2019). The promise and challenges of Quantum Smart Industry 1.0. Quantum Journal, 1(1), 6.

[49] Quantum Computing and Industry. (2020). IBM Research. Retrieved from https://www.ibm.com/quantum/research/industry

[50] Zhang, Y. (2021). Quantum sensing technologies in industry. Quantum Science and Technology, 6(1), 014003.

[51] Wehner, S. (2018). Quantum communication. Science, 362(6414), eaat0347. Biamonte, J. (2017). Quantum Machine Learning. Nature, 549(7671), 195-202.

Index

For Product Safety Concerns and Information please contact our EU
representative GPSR@taylorandfrancis.com
Taylor & Francis Verlag GmbH, Kaufingerstraße 24, 80331 München, Germany

www.ingramcontent.com/pod-product-compliance
Lightning Source LLC
Chambersburg PA
CBHW060351220326
41598CB00023B/2882

9 781032 453637